Musabekova Araylym

CROP PRODUCTION

Musabekova Araylym

CROP PRODUCTION

Textbook

ScienciaScripts

Imprint

Cover image: www.ingimage.com

This book is a translation from the original published under ISBN 978-620-7-48566-6.

Publisher:
Sciencia Scripts
is a trademark of
Dodo Books Indian Ocean Ltd. and OmniScriptum S.R.L publishing group

120 High Road, East Finchley, London, N2 9ED, United Kingdom
Str. Armeneasca 28/1, office 1, Chisinau MD-2012, Republic of Moldova, Europe
Managing Directors: Ieva Konstantinova, Victoria Ursu
info@omniscriptum.com

Printed at: see last page
ISBN: 978-620-8-51731-1

Contents

PhD. Musabekov Aidos Tumenbaevich senior lecturer of the Department of Microbiology and Biotechnology, Deputy Dean for Academic Work of the Faculty of WTW, Kazakh Agrotechnical Research University named after Saken Seyfullin.
D.B.Sc. Abzhalelov Akhan Begmanovich Professor of the Department of Management and Engineering in the field of environmental protection at L.N. Gumilev Eurasian National University.

The textbook describes botanical and biological features of the main field crops, as well as technologies of their cultivation. It is intended for students of agronomic colleges studying in the specialities "Agronomy", "Agrochemistry and Soil Science". It can be used by students of agronomic specialities of higher educational institutions.

INTRODUCTION

Grain production is the basis of the agro-industrial complex. Grain is used for the production of foodstuffs (flour, bread, confectionery, cereals), animal feed, and is a raw material for industry in the production of alcohol, starch and other products. Kazakhstan is one of the largest grain producers in the CIS.

Crop productivity and yield quality depend on many factors. The highest yields and quality are achieved under the combined effect of optimal conditions of plant growth and development, many of which are not regulated, but some of which can be taken into account in practice through the correct choice of technological methods of crop cultivation. In this regard, a clear understanding of environmental factors and conditions, the requirements of plant biology to the factors and the techniques aimed at meeting these requirements in order to obtain high yields of good quality with minimal inputs is necessary.

The purpose of this work is to master the theoretical foundations and practical skills in the selection and implementation of certain technological methods in the cultivation of major crops.

The main attention is paid to the main field crops cultivated in Kazakhstan. In this regard, the specialists' knowledge of biological peculiarities that allow, taking into account the weather conditions of a particular year, to apply technological methods that provide more favourable conditions for growth and development of plants of field crops is of particular importance.

CROP PRODUCTION AS A BRANCH AGRICULTURAL PRODUCTION

One of the most difficult problems of our time is the provision of food for the population. Hunger nowadays is not a spectre, but a real threat facing 1/3 of the world population. According to FAO estimates, in developed countries there are 3000 kcal per 1 inhabitant per day, and in developing countries - 2150 kcal, in the first case - 116% of the norm, in the second - 93%. In addition to apparent hunger, there is so-called hidden hunger, i.e. lack of protein in food, especially of animal origin. In many respects the solution of this problem depends on agriculture, which is represented by two branches - crop production and animal husbandry. Crop production provides 88 per cent of people's needs in food energy (carbohydrates, fats) and 80 per cent in protein. Globally, plant products account for 90 per cent of the human diet.

Crop production is a branch of agriculture, the main task of which is to grow plants for the production of products that satisfy human needs for food, animal feed, raw materials for the processing industry, and in the near future, plant products are an alternative renewable source of energy.

Crop production is the science of cultivated plants, their biological requirements to environmental conditions, methods and techniques to meet these requirements and cost-effective cultivation of maximum yields of high quality. About 20 thousand species of plants are cultivated on the globe. Plant science as a science studies about 90 species belonging to field culture.

The object of study of plant science is plants that create organic matter from inorganic matter (water and CO_2) through the process of photosynthesis using energy from the sun.

The research methods are field, vegetation, laboratory, and production experience.

About the development of plant cultivation and its current state. Man began to cultivate plants 6-8 thousand years ago in Syria, Egypt, India and China. The gathering of wild plants was replaced by primitive farming, when human and animal muscle power was used to propel tools.

In the modern period, the development of crop production followed the extensive path, when the increase in gross output was due to an increase in the area under crops. An example of extensive development is the development of virgin lands, when the total sown area in the USSR increased from 203 to 211.5 million hectares. A large base of grain production in the country was created . Along with this, in the 80's methods of intensive farming - the introduction of scientific achievements, new varieties, high-performance machinery, mineral fertilisers, chemical means of plant protection allowed to increase the yield of grain crops from 10 to 17 c/ha in the USSR. But if earlier a hundredweight yield of 16 centners/ha was considered generous, then by world standards it is now a low level of yield compared to 40-50 centners/ha on average in Western Europe and the USA annually. Achievement of such level of grain crops yield not only in developed but also in some developing countries was achieved as a result of the so-called "green revolution" in the second half of the XX century, when varieties of grain crops, especially responsive to increased doses of mineral fertilisers, short-stemmed, not lodging at a yield of 40-50 c/ha, were developed. Their cultivation required large expenditures of energy to ensure optimal conditions (fertilisers, irrigation, protection means). Tractors of 50 hp were replaced by tractors of 100 hp. If at the beginning of the century doubling of crop yields cost 3-4-fold increase in energy consumption, at the end of the century the same growth requires 5-10-fold increase in energy consumption for production.

If earlier the most abundant yields were obtained on chernozems, now - in Western Europe, where soils are not distinguished by high natural fertility, due to the use of mineral fertilisers, the amount of which has increased dozens of times, and the yield - 2-3 times.

Experts estimate that about 1/3 of the crop is killed by "green fire" - weeds and pests. The cost of defence has increased 20 times, while the yield has increased 3 times.

As a result, it turns out that primitive agriculture of Asian and African countries using manual labour received 10-20 kcal per 1 kcal spent, while the coefficient of energy conversion (conversion) by mechanised agriculture turns out to be 2-3. Further it can be 0.3-0.5, i.e. to produce 1 kcal of food it will be necessary to spend 2-4 kcal of energy.

The main task now is to make agriculture less energy-intensive. The main directions of its development are energy and resource saving, reasonable intensification that does not disturb the balance of the environment, as well as biologisation of agriculture. The main ways of adaptive, biological alternative crop production are the wider use of organic fertilisers, as well as techniques that increase soil fertility. Instead of chemical plant protection products, biological ones are used more intensively. There is a need for a wider spread of crops that maintain and reproduce soil fertility.

1. GENERAL CHARACTERISATION CEREAL CROPS

Grain breads include:

- soft wheat (Triticum aestivum L.), durum wheat (Triticum durum Desf.), barley (Hordeum vulgare L.), oats (Avena sativa L.), rye (Secale cereale L.), triticale (Triticosecale Witt- mach) - breads of Group I;
- maize (Zea mays L.), millet (Panicum miliaceum L.), sorghum (Sorghum vulgare L.), rice (Oryza sativa L.) - breads of group II (Table 1). They belong to the bluegrass family (Poaceae).

Table 1

Generic differences of breads of group I and II

Signs	Bread Group I	Bread Group II
1. Presence of a longitudinal groove on the ventral side of the grain	Available	Absent
2. Number of germinal roots at grain germination	3-8	1
3. Development of flowers in the spikelet	The lower flowers are better developed	The upper flowers are better developed
4. heat demanding	Low	High
5. Moisture requirements	High	Low
6. Availability of winter and spring forms	There are winter and spring crops	Only spring crops are available
7. Development in the initial phases	Quick	Slow
8. Photoperiodism	Long day plants	Short day plants

1.1 Anatomical structure of the granule

The fruit of grain breads is a granule covered with seed and fruit sheaths, which are fused with the seed and constitute 5-7% of the granule weight. In film breads, the stalk is also covered with flower scales. Under the seed coat is the endosperm with a reserve of nutrients (70-85% of the grain weight). The outer layer of endosperm - aleurone - consists of one row of cells (3-5 rows in barley), rich in nitrogenous substances and enzymes that promote grain germination (Fig. 1).

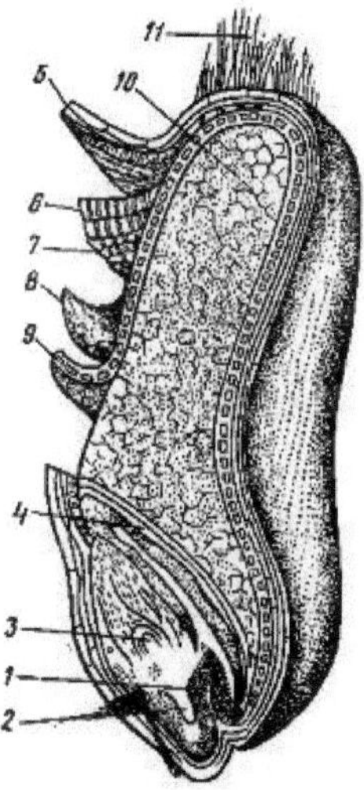

Figure 1. Cross section of a wheat grain:

1 - a foetus;
2 - rudimentary roots;
3 - kidney;
4 - dashboard;
5 and 6 are fetal shells;
7 and 8 are the seed coats;
9 - aleurone layer;
10 - endosperm;
11 - tuft

The inner part of the endosperm is powdery and filled with starch grains, between which protein is deposited. At the base of the grain is the embryo (2-12% of the grain weight). The embryo consists of a germinal root, a stalk and a bud with rudimentary leaves. Between the embryo and the endosperm there is a shield - the only modified seed pod (Fig. 1). Distinctive features of cereal grains are shown in Table 2 and Figure 2.

Distinguishing features of cereal grains

Table 2

Culture	Captivity	Shape	Surface	Colouring	Crest
Breads of group I, there is a groove on the ventral side					
Wheat	Usually naked.	Oval, ovoid	Smooth	White or red	There is or is weakly
Rye	Naked	Elongated, pointed towards the base	Finely wrinkled	Greenish or yellowish	Available
Barley	Filiform, scales fused with grain, less often naked	Elliptical, pointed at the ends	The films show relief longitudinal neuralisation	Yellow or black	Absent
Oats	Filiform, scales not fused with grain, rarely naked	Elongated, tapering towards the top	Smooth in films, without films with hairs	White, yellow, brown	It's well-defined
Breads of group II, there is no groove on the ventral side of the grain					
Maize	Naked	Rounded, faceted	Smooth, wrinkled	White, yellow, miscellaneous	-
Millet	Filiform, the scales are not fused to the grain	Rounded	Smooth, glossy	White, yellow, red	-
Sorghum	Filmy or naked	Rounded	Smooth, shiny	White, yellow, brown, black.	-
Rice	Filamentous, flower scales fused to the grain	Elongate-oval, flattened.	Longitudinally ribbed	Straw yellow, brown	-

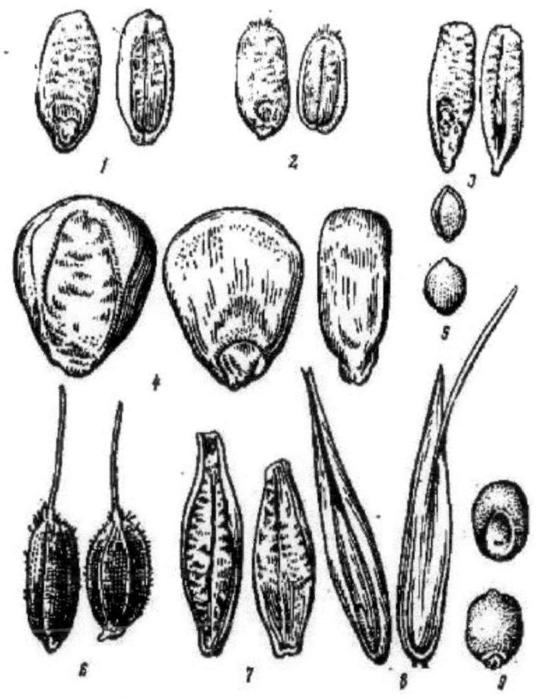

Figure 2. Grains of bread cereals:
1 - durum wheat; 2 - soft wheat; 3 - rye;
4 - maize; 5 - millet; 6 - rice; 7 - barley; 8 - oats;
9 - grain sorghum

1.2 Chemical composition of bread cereal grains

The chemical composition of grain depends on the species and variety of crop, cultivation conditions and agrotechnics (Table 3).

The main content of the granule is carbohydrates (nitrogen-free extractive substances - BEV), which are represented mainly by polysaccharides, and most of them (63-74%) is starch. It is contained in the endosperm in the form of starch grains. The remaining amount of carbohydrates is represented by sugars (2-3%) - mainly in the germ.

Table 3

Chemical composition of bread cereal grains,
% of absolute dry matter (ADM)

Culture	Proteins	Carbohydrates	Fats	Ash	Fibre
Soft wheat	13,9	79,9	2,0	1,9	2,3
Durum wheat	16,0	77,4	2.1	2,0	2,4
Rye	12,8	80,9	2,0	2,1	2,4
Barley	12,2	77,2	2,4	2,9	5,2
Oats	11,7	68,5	6,0	3,4	11,5
Maize	11,6	78,9	5,3	1,5	2,6

Rice	7,6	72,5	2,2	5,9	11,8
Millet	12,1	69,8	4,5	4,3	9,2

The most valuable part of the grain is also represented by proteins. Hard and soft wheat is the most rich in proteins.

Proteins are made up of amino acids. The more essential amino acids (valine, lysine, tryptophan, methionine, leucine, etc.) are present in the protein composition, the more valuable the protein is for food and feed purposes.

Proteins are complex (lipoproteins, nucleoproteins) and simple (albumin - water-soluble proteins, globulins - proteins soluble in weak solutions of neutral salts, gliadins - soluble in 70-80% alcohol, glutenins - soluble in weak solutions of acids and alkalis). The most valuable are glutenins and gliadins. The best ratio for baking is 1:1.

Proteins that are insoluble in water are called gluten proteins. Gluten is a clot of protein substances remaining after dough is washed from starch and other constituents. Flavour and baking qualities of flour depend on the quantity and quality of gluten in the grain. Raw gluten content of wheat is 16-52%, rye - 8-26, barley - 16-20, triticale - 2844%.

Good gluten stretches in length without tearing, resists stretching.

Protein and gluten content in the grain of all breads increases when moving from north to south and from west to east, as climate dryness and nitrogen content in the soil increases. For example, in the Altai Territory in the steppe wheat grain is formed with protein content of 15%, gluten - 32, and in the foothills of Salair - protein 12, gluten - 22%.

The following agrotechnical methods increase protein and gluten content in grain: 1) placement on the best predecessors (fallow, perennial leguminous grasses, leguminous crops); 2) application of nitrogen fertilisers, organic matter, foliar feeding with urea in the phase of earing - grain filling; 3) protection of plants from diseases and pests, especially from the turtle bug; 4) harvesting in the phase of waxy ripeness; 5) prevention of grain germination and frost damage.

The fat content of cereal grains ranges from 2 to 6%. It is most abundant in the germ. If the fat content in the grain is high, e.g. in such crops as maize, millet, oats, flour burns quickly and is poorly stored. For this reason, the germ is removed from maize before grinding , from which maize oil is prepared.

Grain includes ash (from 2 to 5.9 per cent). These are mineral or ash substances (phosphorus, potassium, calcium, sodium, iron, silicon, sulphur, etc.). Vitamins in grains are represented mainly by the B group.

1.3 Structure and development of cereal grains

Yield is formed in the process of plant growth and development. Growth is the addition of dry mass of the plant in the process of assimilation. Development is the formation of specialised organs and plant parts. Grain cereals have the following phenological phases of development, distinguished by the appearance of new organs and a number of external signs: germination, sprouting, tillering, tube emergence, earing (sprouting), flowering, waxy ripeness, full ripeness. The beginning of a phase is marked when 10 per cent of plants are in this phase, and the full phase is marked when 75 per cent of plants are in this phase.

Heat, water, and air are necessary for seed germination. Water is needed for swelling of the grain and enzyme activity. Grains of cereals absorb water in the amount of: wheat - 47%, rye - 58, barley - 48, oats - 60, maize - 44, millet and sorghum - 25% of the grain weight. Larger grains with denser shells and higher protein and fat content swell more slowly.

Under the influence of enzymes in the grain, hydrolysis (i.e. disintegration and dissolution with water absorption) of the reserve nutrients of the endosperm takes place. Proteins are transformed into amino acids, starch into simpler carbohydrates: dextrin, glucose, fructose, which are delivered through the shield to the embryo, and it begins to germinate. Germinal roots are the first to start growing. Breads of the first group germinate with several roots: wheat - 3 to 5, rye - 4, barley - 5-8, oats - 3-4. Breads of group II are germinated by one germinal root.

Temperature affects the speed of water absorption and sprouting appearance. For this period for breads of the first group the minimum temperature is +2...+3oC, optimal - +15...+20, and for breads of the second group the minimum germination temperature is +10, optimal - +25^{0}C.

Sprouting is the appearance of a stem shoot on the soil surface in the form of a spike covered with a transparent modified leaf, called coleoptile, which protects the stem and the first leaf from damage during growth in the soil. Distinctive features of seedlings are shown in Table 4.

Table 4 Distinguishing features of bread cereal seedlings

Culture	Signs of a leaf			
	width	pubescence	colouring	location
1	2	3	4	5
Winter wheat	Narrow	Naked	Emerald green	Vertical to the soil surface
Spring wheat	Narrow	Naked or pubescent	Light green	Same
Rye	Narrow	Naked or faintly	Purple-brown	Same

		pubescent		
Barley	Medium width	Same	Lime green	Same
Oats	Wide	Same	Green or light green	Same
1	2	3	4	5
Maize	Wide, funnel-shaped open.	Same	Green	Slightly bent downwards
Millet	Same	Densely pubescent	Same	Same
Sorghum	Medium width	Bare or sparsely pubescent	Same	Same
Rice	Narrow	Bare, less often pubescent	Same	Vertical to the soil surface

Under optimal conditions, seedlings appear in 4-6 days. A week after the 1st leaf emerges, the 2nd leaf appears, followed by the 3rd leaf at the same interval. At the same time, germinal roots develop to a depth of 30-50 cm. Then comes the tillering phase.

tillering is the formation of additional shoots and adventitious nodal roots from the tillering node located 2-3 cm deep in the soil. A tillering node is a zone of converging stem nodes.

Optimum conditions for tillering are temperature +10...+120C, moistened topsoil, short day. Under favourable conditions, cereals produce 5 to 8 shoots per plant in winter crops and 2 to 3 in spring crops. Barley and oats grow better among spring crops. Productive bushiness increases yield. Productive bushiness is the average number of stems with an ear in which a full grain is formed on one plant. Late tillering results in the formation of podding (lateral shoots with unproductive spikelets) and podsed (shoots without spikelets).

The root system of cereal crops is axillary. First, germinal primary roots appear, they penetrate to a great depth. Then, in the tillering phase, secondary nodal roots appear, developing mainly in the arable horizon. In high-growing cereals (maize, sorghum, millet), aerial or support roots can be formed from above-ground stem nodes. All types of roots are of great importance for plants.

The trumpeting phase is the growth of the stem in height. The stalk of bread cereals is a straw, consisting of 5-7 (in some crops up to 25) internodes separated by stem nodes. The number of internodes is equal to the number of leaves. Straw is hollow in most cereals, but in maize and sorghum it is made of parenchyma.

Stem growth starts with the elongation of the lower internode, then each subsequent internode grows, overtaking the previous one in growth. Each internode grows with its lower part, so the upper part of the

internode becomes harder earlier. This type of growth is called intercalary or insertion growth.
The beginning of the tube emergence phase is marked when a thickening - close stem nodes - is felt above the soil surface at a height of 5 cm inside the leaf sheath. Stem growth in height continues until flowering. As the plant grows, leaves are formed.
The leaf of cereal breads is linear-lanceolate, consisting of a leaf plate and a leaf sheath by means of which it is attached to the stem. At the point of transition of the sheath to the plate, there is a ligula, a thin colourless film that fits tightly to the stem, preventing water from penetrating into the sheath. At the base of the leaf sheath there are auricles (auricula) covering the stem on both sides. Breads of the first group in the phase of tube emergence differ well from each other in the structure of auricles and reed (Fig. 3, Table 5).

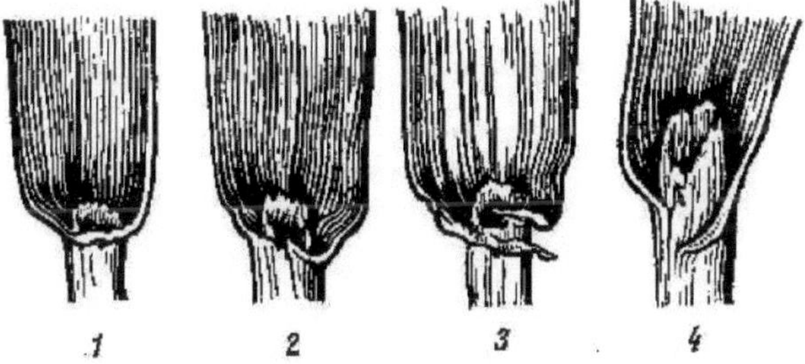

Figure 3. Ears and uvulae of bread cereals:
1 - rye; 2 - wheat; 3 - barley; 4 - oats

Table 5

Distinguishing cereal crops by ears and tongues

Signs	Wheat	Rye	Barley	Oats
Ears	Small, often with cilia	Short, no eyelashes, drying out early.	Very large, no eyelashes, go behind each other	Absent
Tongue	Short	Short	A small one	Large, serrated edges

Spiking (flapping) occurs when the spike or panicle is 1/3 out of the sheath of the upper (flag) leaf of the main stem.
Grain cereals have inflorescences: ear (wheat, barley, rye, triticale) and panicle (oats, millet, sorghum, rice). Maize forms two inflorescences on one plant: cob - female inflorescence and panicle - male inflorescence (Fig. 4, 5, Table 6).
The ear consists of an ear shaft with spikelets on its ledges. One row of spikelets is visible on the front side of the ear and two rows on the side.
The panicle consists of a central axis with side branches of the first order, second order, etc. Spikelets sit on the ends of the branches.

A spike consists of one or more flowers enclosed in spikelet scales. The flower has flower scales (outer or lower and inner or upper). Between the flower scales there is an ovary with a two-lobed pinnate stigma and three stamens (rice has six stamens).

The period "tube emergence - earing" is very important for plants, as the main elements of the ear (panicles) continue to form, further growth of plants takes place, and at this time they are especially demanding to moisture, light, nutrients.

Distinguishing grain breads by their inflorescences

Culture	Inflorescence	Number of spikelets per ear	Number of flowers per spike	Spikelet scales	Outer flower scales	Place of attachment of awns in spinous forms
1	2	3	4	5	6	7
Wheat	Kolos	1	3-5	Broad, multinerved, with longitudinal keel and denticle	Smooth, no keel	To the apex of the outer flower scale
Rye	Kolos	1	2, often with a rudimentary third	Very narrow single-nerve, with a longitudinal keel	With keel, margin ciliate, passing to awns	Same
Barley	Kolos	h (in double-row barley, two of the three are underdeveloped)	1	Very narrow, linear, flat, without keel, with osteo-prominent acuminate at the top	Broad pentanervous, passing into awns in spinous forms	Same
Oats	Broom	1	2-4, rarely 1	Broad, large, with longitudinal nerves, webbed	Smooth, no keel	To the dorsum of the outer floral scale

1	2	3	4	5	6	7
Maize	Male - broom	2	2	Broad, pubescent, with longitudinal nerves	Thin, filmy.	-
	Female - cob	The spikelets are arranged in pairs, in rows	2, fruiting only the upper one	Small, located at the base of the grain	Small, filmy, at the base of the grain	-
Millet	Broom	1	1-2	Rounded-convex, webbed, multinerved two larger, third shorter	Smooth glossy	-
Rice	Broom	A few	1	Narrow, linear lanceolate	Broad, scalloped, pubescent	To the apex of the outer floral scale
Sorghum	Broom	2-3, 1 - fruiting, sessile, sterile on stalks, fall off	1	Convex, leathery, pubescent or glossy	Delicate thin	To the base of the outer flower scale

Figure 4. Inflorescences of maize and panicles of bread cereals:
1 - female inflorescence of maize - cob;
2 - male inflorescence of maize - panicle; 3 - panicle of millet;
4 - rice panicle; 5 - oat panicle

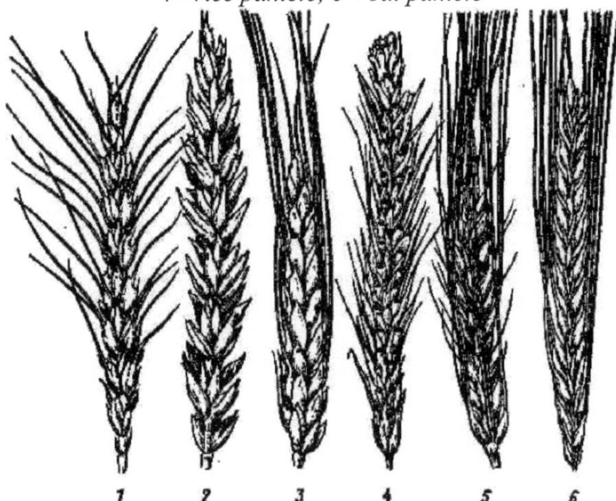

Figure 5. Inflorescences of spikelets of bread cereals:
1 - soft awned wheat; 2 - soft awnless wheat;
3 - durum wheat; 4 - rye; 5 - multi-row barley;
6 - barley double row

Flowering (fertilisation) occurs shortly after earing. In openly flowering plants, the flowers open and the anthers and pistil stigmas emerge. Under unfavourable conditions during flowering (high temperature, drought, frost, rainfall), pollination is incomplete, and as a result, the pollen is observed through the grains.

Grain breads are self-pollinated (wheat, barley, oats, millet, rice) and cross-pollinated (maize, sorghum, rye, triticale) according to the pollination method. In self-pollinators, pollen maturation occurs while the flower is still closed, and the pollen falls and germinates on the stigma of the same flower. Self-pollinators can bloom either with open flower scales (in clear sunny weather) or closed (in cloudy weather), in which case strict self-pollination occurs. Barley is a strict self-pollinator, where pollen is deposited on the flower stigma during earing.

Cross-pollinating plants flower with open flower scales. The ripened anthers crack and the pollen is carried by the wind, germinates on the stigma of the pistils of other plants, and fertilisation takes place.

After flowering, grain formation begins. N.N. Kuleshov divided this process into periods: formation, filling and ripening.

Formation lasts from fertilisation until the final length of the seed is established.

Bulking lasts from the beginning of starch deposition in the endosperm to the end of this process. Bulking takes place in four stages: watery state (2-3% of dry matter accumulates), pre-milk state (10% of dry matter accumulates), milk state (50% of ASW accumulates in the grain, grain moisture content 60%) and doughy state of the grain (85-90% of ASW, moisture content 40%).

Grain ripening begins with the cessation of nutrient supply to the grain. Maturation takes place in two phases:

- waxy **ripeness phase** - endosperm waxy, cut by nail, grain coats yellow, leaves and stems yellow, except for 2-3 upper nodes, grain moisture 30-
35%, lasting 3-6 days.
- **phase of hard ripeness** - endosperm is hard, powdery or vitreous, cannot be cut with a fingernail, shell is dense, leathery, grain colour is typical for the variety, leaves and stems are yellow, grain moisture is 20-15%.

In the phase of full ripeness, as well as for some time after harvesting, post-harvest ripening takes place in the grain, the synthesis of high molecular weight organic substances ends, respiration fades, and full ripeness of the grain occurs, when seed germination reaches its maximum value.

CHAPTER 2

2. *OZYME BREADS*

2.1. Distribution, yield, origin of winter crops

Winter crops are widely spread (Table 7).

Table 7

Sowing area and yield of winter crops

Cultures	Russia		Altai Territory	
	area, mln ha	yield, tonnes/ha	area, thousand hectares	yield, tonnes/ha
Winter wheat	7,7	2,2	21	1,8
Winter rye	4,0	1,5	112	1,7
Winter barley	0,39	2,6	-	-

The area under wheat sowing in Russia is 22.2 million ha, of which about 1/3 (7.7 million ha) is winter wheat. In Russia, winter wheat is mainly distributed in the European part of Russia. Winter barley in Russia is sown in southern regions and is not used at all in Siberia due to its low frost resistance. Winter rye is widespread in the Non-Black Earth Region and Siberia.

Wheat is one of the most ancient crops. 6.5 thousand years ago it was cultivated in Iraq, Egypt and Asia Minor. Barley has the same ancient origin.

According to N.I. Vavilov, wild species of rye served as a basis for the emergence of weed-field rye in South-West Asia, from which the cultivated rye subsequently emerged, and rye heavily contaminated wheat, and with the advancement of wheat crops to the north, rye as more winter-hardy displaced wheat from crops and became an independent crop in the III-IV centuries. Therefore, rye is a relatively young crop. The area under rye cultivation in the world is 11 million ha (Russia, Germany, Poland, France, USA).

In Altai, winter crops are sown mainly in forest-steppe areas of the foothills, where a large snow cover protects winter crops from freezing.

Triticale is a new botanical genus artificially obtained by man through distant hybridisation of wheat and rye, combining valuable properties of these crops (quality of wheat grain and unpretentiousness of rye). As a hybrid crop, triticale is more productive due to its very large grain and high ear fineness, but its quality is somewhat inferior to wheat. The grain can be used both in baking and for livestock feed. The name amphidiploid - wheat-rye hybrid - was derived from the word Triticum (wheat) and Se- cale (rye). V.E. Pisarev created octaploid 56-chromosome triticale from crossing soft wheat and rye. Hexaploid triticale were created from crossing durum wheat and rye, they are more valuable as they have high graininess and high protein content.

According to the type of development there are winter and spring forms of triticale, mainly winter forms are used in production.

Winter crops, as a rule, are more productive than spring crops, as they have a longer assimilation period - 120-150 days, while spring crops have 90-100 days. With good development from autumn, winter plants have a root system up to 1 m, 4-8 stems, so in spring they better use winter-spring moisture reserves and nutrients in the soil, and suffer less from early droughts. Potential yield of winter wheat and triticale is 5-9 tonnes/ha, rye - 4-7 tonnes/ha. Winter wheat accounts for 30% of the gross grain harvest in Russia, but this is not enough. Increasing yields and expanding the area under winter crops is an important reserve for increasing grain production.

Agrotechnical importance of winter crops is that overwintered plants quickly start growing in spring and choke out weeds (especially high-growing rye), so their crops are cleaner than spring crops. Harvesting of winter crops takes place 10-15 days earlier, which allows for more thorough preparation of the harvesting area and reduces the intensity of harvesting operations.

2.2. Differences between winter and spring forms. Hardening of winter crops

Grain breads have two biological forms: winter and spring.

Winter breads include winter wheat, winter rye, winter barley and winter triticale.

In plants, the following stages of development are distinguished:

1) The dark stage takes place from the formation of the zygote on the mother plant until the 11th stage of organogenesis. This stage is characterised by an undifferentiated growth cone;

2) light stage includes further development starting with differentiation of the growth cone. Yarrowisation is the transition from stage I to stage II. Yarrowisation is also the process of accelerating the transition from vegetative to generative development under the influence of temperature or other factors.

The difference between winter and spring biological forms is as follows: winter crops have a longer stage of jarrowisation - 35-60 days. It requires conditions that are usually absent in spring, such as a low temperature from 0 to +10^{0}C. This explains the need to sow winter crops late in the summer, 50 _COPY . This explains the need to sow winter crops in late summer 50- 60 days before the onset of stable cold weather, and the grain harvest is obtained the following year. Otherwise, when sowing in spring, when the period of low temperatures is too short, winter crops do not have time to pass the stage of jarovisation, do not form a stem and generative organs, and do not yield grain.

Spring crops have a short period of yarrowing - 7-20 days, requiring

higher temperatures (+5...+20oC), so they are sown in the spring, and the harvest is obtained in the same year.
There are also intermediate forms - two-stranded forms, which pass the stage of jarovisation at +3... .+15° C and in southern regions can yield seeds both in spring and autumn sowing.
Hardening of winter crops. The growing season of winter crops is 26°-36° days (assimilation period is 12°-15° days). The autumn development period is 5°-55 days. During this time the plants form 3-4 shoots 15-20 cm high, which is the optimum condition for overwintering. In autumn, the plants undergo hardening off. This is a physiological process that results in the formation of properties:
winter hardiness (ability to resist a complex of unfavourable conditions in winter) and frost resistance (resistance to low temperature). Hardening phases: 1) at low temperatures +8...+10oC (at night 0oC) in the light, sucrose up to 30% of ASV is accumulated in the tillering node, as respiration and growth slow down, while assimilation continues. The plant can withstand up to -1°° C; 2) the main phase, when dehydration of cells, outflow of water from cells into intercellular spaces, conversion of insoluble nutrients into soluble ones takes place. An increase in the concentration of cell sap causes its freezing point to decrease. The phase takes place both in the light and in the dark, at a temperature of 0...-5oC. Hardening lasts 20-25 days. It is better that it takes place in clear weather with warm days and cool nights. Optimal sowing dates are also important, as the plants should have time to accumulate a stock of nutrients. Positively affect the process of hardening, increase winter hardiness of winter additional phosphorus and potassium, as they contribute to a faster formation of roots and accumulation of carbohydrates in the node tillering. Nitrogen, introduced in autumn, has a negative effect, as it increases the growth of above-ground mass, thereby reducing the stock of sugars in the tillering node. Varieties with a longer period of yarrowing, which spend less nutrients on respiration when the heat returns in autumn, as well as varieties with a deeper tillering node, accumulating more oligosaccharides, proline, asparagine and glutamic amino acids, are more winter-hardy.

2.3. Causes of winter crop deaths and prevention measures

Unfavourable conditions in autumn, winter, early spring can lead to the death of winter crops. There can be various reasons for this.
1. Frost is one of the most frequent causes, especially in areas with severe winters. Under the action of prolonged frosts, water freezes in the intercellular spaces, draws water away from the cells, dehydration, colloid coagulation, disruption of cytoplasm structure and cell death occur. The deeper the tillering node, the more winter-hardy plant is

formed, so deeper but optimal sowing in terms of depth and timing contributes to better preservation of plants. With furrow sowing, the tillering node is formed deeper, as the light signal to the tillering zone arrives faster. Snow retention is of great importance. It is better to do it with the help of wings. At frost -30^{0}C under a layer of snow 15 cm the temperature in the ground layer -11oC, and under a layer of 50 cm -2 ... -3oC, as snow has low thermal conductivity. Winter rye and triticale are more frost-resistant, withstanding -20^{0}C at the tillering node, while wheat -16... -180C. -18oC.

2. Hardening is observed when a large layer of snow falls on unfrozen soil. The temperature under it remains close to zero, plants continue to breathe, consume plastic substances, and become exhausted, as assimilation without access to light does not take place. The death of plants occurs in 90-100 days at a temperature of 0^{0}C. Rolling in autumn after snowfall prevents uprooting, as it promotes rapid freezing of snow and soil. Plants at negative temperatures stop breathing intensively and fall into anabiosis. Often overgrown winter crops from autumn also sprout out. This is especially true for winter rye, as it develops faster in autumn. Overgrown winter plants from autumn should therefore be mowed. In spring, plants can also dry out if there is a lot of snow and it does not melt for a long time, especially in depressions. Spreading ash, peat, mineral fertilisers accelerates snow melting and prevents the plants from sprouting.

3. Soaking is observed in depressions under melt water. As a result of anaerobic processes, plant death occurs within 12-15 days. It is necessary to make open drainage (furrows), levelling, field leveling, ridge sowing, pre-harvest soil loosening.

4. Sticking out is the displacement of the tillering node to the soil surface when the soil settles during alternating thawing and freezing. Root rupture caused by subsurface cellular ice also occurs. Ice is known to expand when it freezes and tears tissue. Bulging occurs more often in heavy soils. This can be prevented by sowing varieties with a deep tillering node. The last deep tillage should be done a month before sowing, so that the soil has time to settle. Rolling before and after sowing is obligatory. Seed treatment with retardants promotes deeper rooting of the tillering node. In spring, if the seedlings are bulging, the soil should be compacted, then the tillering nodes and roots are pressed to the soil and new secondary roots and shoots are formed.

5. Snow mould is caused by the fungus Fusarium nivale and produces white or pinkish patches on plants. Often snow mould develops on plants weakened by desiccation. Sclerotinia sclerotinia, a white blight with black dots, is also observed for the same reason. Control measures: resistant varieties, seed treatment (Maxim KE, Fundozol), snow

removal, harrowing in spring to remove diseased plants.
6. The cause of winter crops death may be ice crusts, and the lapped crusts are more dangerous than hanging crusts, as they completely freeze with the soil. Control measures: slitting in late autumn, spreading mineral fertilizers, ash, peat.
7. Return of cold weather in spring (-16...-200C), when there is no more snow. Plants begin to breathe, grow, lose winter hardiness.
8. When sown early from autumn, it is killed by Swedish and Hessian fly.
9. Blowing in dry autumn. Control measures: forest amelioration. strip cropping. flat-cutting tillage.
Early in spring after the snow melt, the condition of winter crops is assessed. Weakly thinned crops with the loss of no more than 15-20% of plants are left to vegetate. On medium-sparse crops with a density of 150-200 plants per 1 m^2 make reseeding, repairs, fertilisers. Strongly thinned crops with a density of 120 plants/m^2 and less are sown with barley or late spring crops.

2.4 Biological characteristics of winter crops

Temperature requirements. The minimum germination temperature is +1.+2^{0}C. The optimum temperature is +15^{0}C. Then sprouts appear in 7 days. In spring, the resumption of vegetation occurs at +5^{0}C. The optimal temperature for tillering is
I 10. I 12, for earing - flowering - +16...+22, for filling and ripening - +25, at +35^{0}C assimilation slows down. The total sum of active temperatures is 1850-2200^{0}C, the vegetation period is 260-360 days.
Rye is more winter- and frost-resistant than wheat. At the depth of the tillering node, rye and triticale can withstand -18.-20^{0}C, winter wheat - 16.-18$^{(0)C}$. Winter barley barely withstands -14^{0}C, and in Siberia it is not sown at all, as it freezes. The distribution of wheat and barley in the country is more southern, rye - more northern.
Winter crops use autumn and winter precipitation better than spring crops, but consume more moisture than spring crops. These plants have well-developed root systems up to 1.5-2 metres. Rye has a stronger root system and is therefore more drought tolerant. The transpiration coefficient is 340-420 for rye and 400-500 for wheat. The critical period for winter crops in terms of moisture is tube emergence - earing, when 60-70% of all moisture is consumed during the growing season.
Soil requirements. Winter rye is less demanding to soil than other grain crops, grows well on sod-podzolic soils. According to D.N. Pryanishnikov, rye is able to convert phosphorus compounds unavailable for other crops into available ones, tolerates high acidity pH = 5.3, as well as loose sandy soils where wheat cannot grow. Light soils with low water holding capacity are more suitable for rye than heavy,

waterlogged, clayey soils. Wheat, like barley, needs more fertile soils - chernozems, chestnut soils with neutral pH not less than 6, with humus content not less than 2%. Triticale as well as rye is less demanding, can grow on grey forest, light loamy soils with pH = 5,5-7,0.

2.5 Cultivation technology

Place in crop rotation. The best precursor for winter crops is a clean tillage fallow. 60% of winter crops in the region are sown on fallow. In humid areas it is possible to use occupied or sidedress fallows. Winter crops can be sown after perennial grasses, legumes and other early-harvested crops, but in this case they winter worse, as they lack moisture and nutrients for development. Wheat and rye are considered to be crops that respond to crop rotation, i.e. using fertilisers and protection means, they can be sown repeatedly. Winter crops, in turn, are good predecessors for row crops and spring crops, as they leave fields clean of weeds.

Soil cultivation. In winter-sown areas of the region the advantage is on the side of black fallow, preparation of which begins with stubble husking after harvesting the crop preceding the fallow by 6-8 cm LDG-10 with an angle of attack 30-35^{0}C, which prevents excessive evaporation, promotes germination of weed seeds, destruction of pests and diseases. Zyablivaya ploughing at 25-27 cm under the steam is done 1.5-2 weeks after husking, after the emergence of weed sprouts. Deep ploughing achieves destruction of perennial weeds, accumulation of moisture and its preservation during the fallow period. Early spring harrowing (ESS) destroys the crust, promotes weed germination, prevents evaporation, improves air and heat regime of the soil. In summer, as weeds emerge, surface cultivation is carried out - cultivation for 6-8-10 cm.

For snow accumulation in the fields, sow waders across the prevailing winds, and on slopes - across the slopes. Coulees are 2-3 rows with row spacing of 60 cm, with a distance between coulees of 10-12 m. Culis seeders SKN-3 or SZS-2,1 are used. Mustard (500 g/ha) and sunflower in arid areas are sown as tiller crops. Sowing date: sunflower - I decade of July, mustard - II decade of July, so that the plants do not have time to be inseminated. It is possible to sow oilseed radish at the same time with winter crops, using seeder SZT-3,6 from grass box with extreme open coulters. If a sideral fallow is used, instead of coulishes sow in the spring siderates (rape, mustard, melilot, millet, sudanka), which are crushed in mid-summer (KIR-1,5, E-261) for mulch, leaving the coulishes of sideral plants. Then disc the soil with BDT-7 no later than 25 days before sowing winter crops.

Deep tillage of the soil for winter crops is prohibited after non-stubble forerunners, otherwise winter crops die in large voids of the loose layer due to root drying, protrusion of tillering node. After stubble non fallow predecessors it is better to cultivate the soil at 8-12 cm (LDG-10, KPSh-9) or sow directly with seeder SZS-2,1 on clean fields. After the first cutting of perennial grasses no later than a month before sowing discing BDT-7, then ploughing at 25-27 cm with harrowing and rolling before and after sowing. Pre-sowing cultivation is carried out to the sowing depth by KPSh-9, KPS-4 with harrows, plumes, rollers at an angle to the main cultivation.

Application of fertilisers. Output for formation of 1 kg of grain and corresponding by-products, kg:

	N	P_2O_5	K_2O
Winter wheat	3,5-3,7	1,3	2,3
Winter barley	3,2-3,6	1,2	2-2,4
Winter rye	2,5-3,5	1,2-1,4	2,4-2,6
Winter triticale	4-5	1,3-1,6	3,6-3,4

Wheat is the most nutrient-demanding crop and is also the most responsive to fertiliser application. 1 kg of fertiliser gives an increase of 8-10 kg of grain. It is recommended to apply 20 tonnes/ha of organic matter under the main tillage (up to 30-40 tonnes/ha in humid areas). If the fallow was filled with organic matter, mineral fertilisers are applied in the rows at sowing, using superphosphate 10-20 d.v. kg / ha, and on soils poor in nitrogen, it is better to use complex fertilisers $N_{15}P_{15}$ kg d.v. / ha. If organic matter was not applied, the main fertiliser is applied in the fallow $P_{40-60}K_{40-60}$ kg d.v. / ha, on other predecessors - $N_{60}P_{60}K_{60}$. If economically feasible, the application rate is brought up to 120 kg d.v/ha for each element. Large norms of nitrogen is better to apply fractionally in the form of fertilisers, otherwise it is washed out and unproductively spent, plants overgrow in the autumn and uproot. Fertilisers applied during critical periods of nitrogen consumption:

1) autumn top dressing is effective only on nitrogen-poor soils, if the main fertiliser has not been applied, as well as on fallow predecessors, in an amount of up to 20% of the main fertiliser;

2) spring nitrogen fertilisation is particularly effective as the soil is depleted of nitrogen at this time. Apply 30-50 kg of nitrogen fertiliser/ha by spreading ammonium nitrate with RMG-4 and then harrowing the soil. It is more effective to use for this purpose disc fertiliser seeders SZ-3.6, SU-2.4 in combination with harrows across the rows. Yield increase can be up to 0.3-0.5 tonnes/ha. Such fertilisation is especially recommended for winter crops thinned after overwintering, as it causes additional tillering and makes up for the lack of plant density;

3) in the phase of tube emergence by air or along the technological

track, feeding with RMG-4 or OPSh-5 (if the fertiliser is in liquid form) increases ear fineness without increasing straw yield;

4) at the earing phase, foliar nitrogen fertilisation with 30 kg d.w./ha does not significantly increase yield, but increases protein and gluten by 1 and 4%, respectively. It is better to use a 30% urea solution as it does not cause leaf scorch. 65 kg of urea is mixed with 150 litres of water, adding trace elements. Cheaper preparations can be used for fertilising, for example, 50 g/ha of sodium humate with the same amount of water. It is possible to apply a solution of float (45 kg of urea + 22 kg of ammonium nitrate + 100 l of water), consumption 100 l/ha, it is better in the morning or evening at a temperature not higher than +20^{0}C, with wind speed not more than 4 m/sec.

Fertilisation is carried out taking into account the diagnostics. *Soil diagnostics*: nitrate nitrogen is determined in soil samples, and if nitrogen reserves are less than 30 kg/ha, it is necessary to apply 20% of the calculated nitrogen rate under main tillage or under seedbed cultivation. If more, nitrogen is not applied to avoid winter hardiness reduction and overgrowth.

Tissue diagnosis: during the emergence phase, nitrate content in plants is determined using a rapid laboratory. Cut 100 plants in the morning, squeeze the juice from the stems onto a glass and apply a drop of 1% diphenylamine solution. White-pink colouration (below 3 points) means that strong wheat cannot be obtained, nitrogen fertilisation is not advisable. Pink colour (3.5-4.5 points) - two fertilisers are recommended at earing and filling, intensely pink - (4.55.5 points) - one fertiliser at earing, crimson - not necessary

feeding, the grain will be strong, but there is a danger of lodging.

Sowing. Winter crops should be sown with seeds from the carryover fund. The carry-over fund is a stock of winter crops seeds made in the previous year for sowing the following year.

Freshly harvested seeds cannot be used to sow winter crops, because they have a reduced germination rate as they have not had time to undergo post-harvest ripening.

When cultivating crops under intensive technology, seeds should be large, with a weight of 1000 seeds of at least 38 g, with a growth force of at least 80%. Before sowing, seeds are warmed in the sun or in dryers, especially in case of sowing with freshly harvested seeds in case of production necessity. Seed dressing is obligatory.

Sowing dates. Plants are better preserved in winter if they overwinter in the tillering phase (3-4 shoots). This requires 50-55 days of autumn vegetation with the sum of active temperatures for rye 400-450^{0}C, for winter wheat - 550-580^{0}C. From the average annual date of transition through + back count 50-55 days - this will be the sowing date of winter crop. Optimal calendar terms of sowing winter crops - from 10 to 25 August. First of all sow on occupied fallows and non-fallow predecessors, then on fallows. More extended sowing dates are possible for rye than for wheat and triticale.

Seeding rates. In the steppe - 3 million/ha, Priobie - 4-5, foothills - 5-6 million/ha of germinated seeds. Under more favourable conditions (on clean fields, on the best predecessors, on fertilized background) one should stick to the lower limit of the recommended rate.

Winter wheat establishes the tillering node somewhat deeper than rye, so wheat is sown at 5-6 cm and rye at 4-5 cm. When the soil dries out, wheat is sown at 7-8 cm, rye at 6-7 cm, and on heavy soils at 3-4 and 2-3 cm, respectively. Triticale is sown at 4-8 cm.

Plant care. After analysing the condition of the overwintered plants, the question of plant care is decided. In case of early warm spring there is an early ecological effect of the time of spring resumption of winter vegetation, in which the snow comes down quickly, plants are well preserved, quickly start to grow, we can expect a high yield, but with lower grain quality, so it will be effective early spring harrowing of crops, the use of drugs - growth regulators to prevent lodging, later nitrogen fertilisers (in the phase of earing - pouring) to improve grain quality.

In years with a late ecological effect of the time of spring resumption of vegetation, plants are often thinned, low-growing, quickly move to earing, form a lower yield, but with a higher protein content. In such years, it is better not to harrow crops in spring; preparations used to prevent lodging are not effective. Early nitrogen fertilisation in the tillering phase or in the tube emergence phase will be effective in increasing the yield by increasing the number of spikelets and flowers in the spikelet.

CHAPTER 3

3. *YARROW WHEAT*

3.1. Meaning and use

Wheat is the most valuable and widespread food grain crop on the globe. Chemical composition of wheat grain, % of ASW: protein - 14 - in soft wheat grain, 16 - in durum wheat grain, carbohydrates - 77-79, fats - 2, ash - 2, fibre - 2.3.

The main product obtained from wheat grain is bread, which has good taste, nutrition and digestibility. Durum wheat is more valuable for use in cereal, pasta and confectionery production. Wheat is used to produce alcohol, starch, gluten, dextrin and glue. Wheat grain and bran is a valuable concentrated feed for farm animals.

The most important indicator characterising wheat quality is protein and gluten content. The protein content is greatly influenced by climate, soil and fertiliser application. The protein content of wheat determines its utilisation. Baking requires grain with a protein content of 14-15%, while pasta production requires 17-18%. The protein digestibility of wheat bread is about 95%.

Strong and hard wheat are of particular value to the flour milling, baking and export industries. Strong wheat is only available in soft wheat. They are characterised by an increased content of protein, gluten and other valuable substances. When assessing the strength of wheat, the baking qualities are decisive. There are 3 groups according to the technological properties of the grain: strong, medium and weak.

Strong wheat is characterised by a higher content of protein in the grain - not less than 14%, crude gluten - not less than 28%, gluten quality - not lower than group I, volumetric yield of bread from 100 g of flour - 550 cm^3, grain vitreousness in red-grained wheat - not lower than 75%, in white-grained wheat - not lower than 60%, baking power of flour - not lower than 280 Joules. Strong wheat is called an improver because of its ability to improve the baking qualities of weak wheat. When flour from strong wheat grain is added to weak wheat flour, the quality of bread is significantly improved. Strong wheat grain is highly valued in the international market.

Medium strength wheat (fillet) has good baking qualities, is able to produce bread of satisfactory quality without adding stronger wheat, but cannot improve weak wheat. Medium wheat grain contains 11-13.9% protein, 25-27% gluten, quality gluten belongs to group II, baking power of flour is 200-280 Jg.

Weak wheat has little baking power. Bread is of low volume and spreads out on the pillow. Weak wheat grain is characterised by lower protein content - less than 11%, crude gluten - 25%, gluten quality - II-III group, volume yield of bread from 100 g of flour - less than 400 cm^3, baking power of flour - less than 200 J. To produce standard bread from weak wheat grain or flour, strong wheat grain or flour is added to it. Wheat is a plastic crop adapted to cultivation on different types of soils and in different climatic conditions. Currently, wheat occupies 213.6 million hectares in the world, which is about 29 per cent of all cereal crops. The largest wheat producing countries are Russia - 19.9 million ha; PRC - 26.6; India - 26.7; USA - 21.5; Canada - 11.0; Australia - 12.1 million ha. The average grain yield is 2.7 tonnes/ha. In Altai Krai, the average area planted is 2.6 million hectares, with an average yield of 1.21 tonnes per hectare for 2001-2004, and up to 3.5 tonnes per hectare at variety plots.

3.2. Wheat classification

Wheat belongs to the family *Roaceae* (bluegrasses), genus *Triticum L.,* which includes 22 species that are well distinguished by morphological and biological features. For practical purposes, all wheat species can be conveniently grouped into two groups: holograin and spelt wheat.

Holosere wheat has an unbreakable ear shaft, the ear does not split into individual spikelets after maturity. The grain is free of spikelet and flower scales during threshing.

The following species belong to the group of holocereal wheat: *Tr. aestivum L.,* (soft wheat); *Tr. durum Desf.* (*durum* wheat); *Tr. carthlicum Nevski.* (carthaline wheat); *Tr. poloni- cum L.* (polonicum wheat); *Tr. turgidum L.* (t*urgidum* wheat); *Tr. turanicum Jakubz.* (Turanian wheat); *Tr. compactum Host,* (dwarf wheat); *Tr. sphaerococcum Per s.* (round-grained wheat); *Tr. amplissifolium Zhuk.* (broad-leaved wheat) and *Tr. fungicidum Zhuk.* (fungicide wheat).

Bast (film) wheat has a brittle ear stem and an ear that easily breaks into individual spikelets at maturity. The grain remains in the spikelets during threshing and special equipment is required to separate it. The other 11 wheat species belong to the spelt wheat group: *Tr. ae-gilopoides Link* (wild single grain), *Tr. urartu Thum.* (Urartu wheat), *Tr. monococcum L.* (cultivated single grain), *Tr. ararati- cum Jakubz.* (Halda wheat), *Tr. dicoccoides Aar.* (wild double grain), *Tr. timopheevi Zhuk.* (zanduri), *Tr. palaeo-colchicum Men.* (Colchicum *dicoccum*), *Tr. dicoccum Schubl.* (emmer, spelt), *Tr. macha Dek. et Men.* (Macha wheat), *Tr. acthiopicum Jakubz.* (Abyssinian wheat) (Fig. 6).

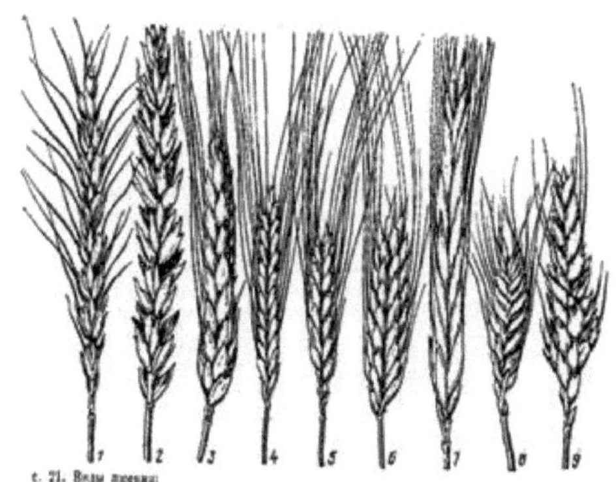

Figure 6. Wheat species:
1 - soft spinous; 2 - soft spineless; 3 - hard;
4 - cultivated single grain; 5 - double grain (spelt);
6- Timofeev wheat; 7- polonikum; 8- dwarf wheat;
9 - turgidum

Soft and durum wheat are of production importance (Table 8). Soft wheat is widely used in baking, so it has the greatest distribution, and durum wheat is used for pasta production, so it has a limited distribution mainly in the more warm steppe regions.

Table 8

Distinguishing features of hard and soft wheat

sign	Soft wheat	Durum wheat
Kolos		
Ear density	Loose, a gap between the spikelets	Dense, no gaps between the spikelets
Widest side	Face	Side
Osti	Equal to or shorter than the ear, diverging	Longer than the ear, parallel
Spike scales	Depressed at the base, with a weakly expressed keel and more or less long denticle	Without indentation at the base, with sharply prominent keel and short denticle
Grain		
Shape	Short, rounded	Oblong, more faceted in transverse section
magnitude	Fine, medium coarse, coarse	Medium, more often large
Consistency	Floury to varying degrees, complete vitreousness is almost never observed	Vitreous, less often lustelike
Fetus	Rounded, wide, concave	Oblong, convex, well defined
Crest	Clearly defined, long hairs	Absent or weakly expressed, hairs short
Solomina	Polaya	Under the spike

White-grain varieties are more susceptible to injury during threshing than red-grain varieties because they have thinner shells. Red-grained varieties have a longer post-harvest ripening period compared to white-grained varieties, which means they are more resistant to germination in windrows and on the root. The awned forms become more important in dry conditions. The awns in this case fulfil the function of additional transpiration in plants under conditions of increased temperature. Even in awnless forms, awns are more pronounced when growing under extreme conditions. But from the economic point of view, the strawless varieties are more preferable, as the straw is better eaten by animals, and the awns do not adversely affect the work of threshing and cleaning organs of the combine harvester. Therefore, red-grained bezeless varieties of soft wheat are more widespread in our conditions. Hard varieties are represented by Gordeiforme.

In turn, species are divided into varieties (Table 9).

Table 9 Varieties of soft and durum wheat

Variety	Presence of awns and their colouring	Spike colouring	Drooping of spikelet scales	Grain colouring
		Soft wheat		
Albidum	Shoeless	White	Unfluffed	White
Lutescence	Same	Same	Same	Red
Alborubrum	Same	Red	Same	White
Milturum	Same	Same	Same	Red
Pyrotrix	Same	Same	Lowered	Same
Velutinum	Same	White	Same	Same
Grekum	The awns are white	White	Unfluffed	White
Erythrosper- mum	Same	Same	Same	Red
Ferruginum	The awns are red	Red	Same	Red
G ostianum	The awns are white	White	Lowered	Red
Barbarossa	The awns are red	Red	Lowered	Red
Cesium	Same	Grey-skinned	Unfluffed	Red
		Durum wheat		
Hordeiforme	The awns are red	Red	Unfluffed	White
Melianopus	The awns are black	White	Lowered	White

3.3. Biological characterisation

Wheat is a long-day plant, rather cold-resistant. During the vegetation process, wheat distinguishes the following phenological phases: 1) seed swelling; 2) seed germination; 3) sprouting; 4) appearance of the third leaf; 5) tillering;

6) tube emergence; 7) earing; 8) flowering; 9) milk ripeness; 10) waxy ripeness; 11) full ripeness.

In each of these phases, plants undergo certain stages of organogenesis and the formation of various elements of yield structure, so it is important to create optimal conditions for each phase. The sprouting phase is preceded by seed swelling, accompanied by water absorption of 45-52% of the seed weight. The glassy, larger seeds of durum wheat require 5-7 per cent more water for germination than the powdery seeds of soft wheat, so the germination of durum wheat is later. The optimum temperature for germination is +12...+150c, the minimum temperature is +20C. +20, to obtain

The temperature +4...+5^{0} is necessary for viable sprouts, so sowing can

be started when the soil warms up to +5^{0}C in order to effectively utilise the spring moisture reserve.
Wheat sprouts are resistant to frosts. In the phase of sprouting plants can withstand up to -4...-50C. In the phase of 3 leaves - tillering - up to -8$^{(0)C}$. Durum wheat has lower resistance to frosts.
In 10-15 days after sprouting, tillering begins - the formation of nodal roots and lateral shoots from the tillering node (zone of close underground stem nodes). Optimum conditions for tillering are +10.+12^{0}C, sufficient moisture, shorter day, therefore in the steppe plants are less bushy, productive bushiness - about 1, the harvest is formed mainly at the expense of primary roots. In more humid conditions with early sowing, the role of tillering in yield formation increases, the number of nodal roots increases, and productive bushiness increases up to 1.53.
The optimum temperature is +18...+200C during the tube emergence - flowering phase. High temperature +38...+400C for 10-17 hours causes paralysis of wheat stomata and formation of sterile pollen. The tube emergence - earing is a critical period in terms of moisture, because at this time, along with the fact that the stem continues to grow in height, leaf surface grows, reproductive organs (spikelets, flowers) are formed in embryonic state, i.e. the potential productivity of the ear is laid down. Water consumption by growth phases is uneven: sprouting - 57%, tillering - 15-20, tube emergence - earing 50-6°, milk ripeness - 2°-30, waxy ripeness 3-5%. Lack of moisture and nutrients in the tube emergence - earing phase leads to the formation of fewer spikelets and flowers. Drought and high temperatures in subsequent earing and flowering lead to disruption of pollination and fertilisation, formation of infertile spikelets, through-grain, hollow spikelets. Optimal temperature during the period of formation and grain filling +22...+250C. High temperature +30...+400C and the presence of dry winds leads to a decrease in yield, the formation of puny grain. In the ripening phase it is favourable to increase the temperature to +25...+280C, which has a positive effect on quality (protein and gluten content, seed germination). During the filling period durum wheat tolerates high temperatures better than soft wheat, better resists dry weather, but in case of lack of heat it has weaker outflow of plastic substances into the grain, and in cold weather its ripening is delayed more. In the filling and milk ripening phase, early autumn frosts lead to the formation of frost-bitten grain with poor baking qualities and low germination.
Good growth of spring wheat requires a certain sum of active temperatures during the growing season equal to 142°-179°° C.
Depending on the soil and geographical zone and weather conditions of

the year in Altai Krai, the vegetation period of early maturing varieties ranges from 75 to 85 days, medium maturing varieties from 85 to 105 days, and late maturing varieties from 105 to 117 days.

Wheat is demanding of readily available nutrients in the soil due to the relatively short period of active nutrient uptake and the reduced digestive capacity of the root system. Hard wheat has the highest requirements to soil fertility, cleanliness and structure, and it succeeds better on chernozem and chestnut soils; soft wheat is particularly favoured by all types of chernozems, chestnut, medium and weakly podzolic soils. On sod-podzolic soils it is necessary to apply lime, organic and mineral fertilisers. Relief conditions have a great influence on wheat yield. Lowered marshy places are also unfavourable for it. Wheat does not tolerate acidic soils. High yields of spring wheat can be obtained in conditions of neutral or slightly acidic reaction (pH_{sol} = 6.0-7.5). Wheat in the initial period of growth is strongly suppressed by weeds, so it is of great importance to sow it on fields clean of weeds. On light soils it succeeds well in more humidified areas. Soft wheat is less sensitive to soil salinity than hard wheat. In general, soft wheat can be characterised as a more plastic crop.

One of the main factors determining the value of yield is variety. A rational combination of varieties differing in vegetation period allows for more efficient use of climatic and material resources and more stable yields from year to year. Spring wheat has varieties with different types of drought tolerance. Early maturing varieties better tolerate late drought, and later maturing varieties better tolerate early drought. Therefore, in the arid steppe in Altai, where the frost-free period is longer and early type of drought is more often observed, it is preferable to combine medium-late and medium-maturing varieties in the ratio of sowing areas 1.5:1. In the moderately arid steppe and forest-steppe of Priobie these same varieties in the ratio of 1:1.5, and in the forest-steppe of the Salair and Altai foothills medium-early and medium-maturing varieties in the ratio of 2.3:1. In steppe areas, varieties with a prolonged sprouting - earing period use summer precipitation more effectively, and in the foothills, medium-early varieties ripen faster in the warmer period of early August and give better quality grain. The combination of varieties with different vegetative periods allows sowing and harvesting to be carried out in an optimally short period of time.

When using different varieties, it is necessary to have information on whether the variety is plastic or, on the contrary, more intensive. In Altai conditions, intensive wheat varieties have not been widely spread, because, being short-stalked, they are less drought-resistant due to poor development of the primary root system, while their secondary root system, on the contrary, is very well developed, which increases their resistance to lodging. But in drought conditions, these varieties do not have the opportunity to show their high potential and are inferior to semi-intensive varieties, which are more plastic, with a height of at least 80 cm and well-branched.

More plastic varieties can be grown on a less favourable background, they reduce yields less, and more intensive varieties can be grown using intensification methods, under fallow, which will allow to obtain higher yield increases.

3.4. Cultivation technology

Place in crop rotation. Wheat has high requirements to predecessors. In the zone of insufficient and unstable humidification the best predecessor is black fallow, which provides accumulation and preservation of moisture, weed control, increases the content of nutrients in the soil. Black fallow not only contributes to the increase of wheat yield, but also makes it possible to obtain high-quality grain meeting the standards for strong wheat when sowing appropriate varieties.

In acutely arid areas, coulis pairs are of great importance. For example, sowing coulis in a pair at the Kulunda Agricultural Experimental Station due to snow accumulation increased the moisture content of the metre layer to 150 mm, which, together with the application of fertilizers, increased yields by 0.47 t/ha.

At present, the application of organic matter to fields is complicated by high transport costs. Therefore, it is more economically favourable to grow siderate. Sideral fallows are especially effective in more humidified zones. According to the data of the AGAU, melilot in a sideral fallow increases wheat yield by 0.20-0.25 tonnes/ha compared to black fallow. In addition, melilot cleans the soil from root rots, wireworms, grain nematodes. This creates conditions for soil erosion protection.

In more humid areas the importance of perennial grasses and peas as forerunners increases, after which wheat gives better quality grain. Spring wheat is also sown after maize and potatoes. Experiments have established that it is inexpedient to sow spike bread on the same field more than two years in a row because of the large accumulation of pests, in particular beetles. Durum wheat is more demanding to predecessors and cultivation technology, as it is a more intensive crop. It is usually grown on fallow or perennial grasses. The positive biological properties of durum wheat include less pest and disease infestation and resistance to shattering.

Fertilisation. Wheat has high requirements for soil fertility and is very responsive to fertiliser. On average spring wheat uses 38-42 kg of nitrogen, 11-12 kg of phosphorus and 25-26 kg of potassium to produce 1 tonne of grain and the corresponding amount of straw. During the sprouting and tillering period, wheat is very sensitive to phosphorus deficiency, which, as a rule, is in a form that is difficult for plants to access. During this period, phosphorus promotes rapid root growth, thereby increasing drought resistance, so it should be applied to the rows at sowing in the amount of 10-15 kg d.w/ha. Phosphorus supply continues until maturity.

Nitrogen is especially necessary during the tillering period, it increases bushiness, the number of flowers and spikelets in the embryonic ear. When foliar nitrogen fertilisers are applied in later phases (earing - bulking), nitrogen is stored in the grain, increasing its quality, which is especially important when growing strong wheat. Potassium is necessary for plants from tillering to grain filling.

In dry steppe on chestnut soils it is advisable to apply small doses of fertilisers: organic matter 10 t/ha, $N_{20}P_{20}$. On southern chernozems of arid steppe and leached chernozems of moderately arid steppe, higher doses of mineral fertilisers $N_{40-60}P_{40-60}$, organics - 10 t/ha are more effective. On leached and podzolised chernozems, grey forest soils of forest-steppe and foothills of Salair and Altai - $N_{60}P_{60}K_{60}$, organics - 20 t/ha. Increases in grain yield when using optimum doses of fertilisers in steppe areas are 0.15 t/ha, in areas of columnar steppe - 0.175 t/ha, in forest-steppe and foothills - 0.20-0.24 t/ha. In all soil-climatic zones at application of nitrogen-phosphorus and complete mineral fertiliser at 40-60 kg/ha the content of gluten in grain increases by 3,0-3,5%, and protein - by 1,0-1,6%. Mineral fertilisers are applied during the main tillage to a depth of 12-14 cm GUN-4, KPG-2,2 or under pre-sowing cultivation SZS-2,1, organic matter is applied in the fallow field under the main tillage, using ROU-6, PT-16.

Tillage. The main task of tillage for wheat in the main areas of its cultivation is accumulation and preservation of moisture and destruction of weed vegetation. The anti-erosion no-tillage system of tillage, when 80% of stubble remains on the surface, allows to accumulate more snow on the fields, protects the soil from wind and water erosion. According to the Kulunda Experimental Station, in all years the yield of spring wheat was 0.16 tonnes/ha higher with flat-cut tillage as compared to mouldboard tillage, and 0.4 tonnes/ha higher in dry years.

In the steppe, immediately after harvesting the predecessor, shallow shallow surface tillage up to 12 cm is made using KPSh-9, OPT-3-5, BMSh-15. Significantly reduces moisture losses in autumn and spring by mulching the soil with straw when harvesting grain predecessor with PUN-5 choppers or by chopping and spreading straw from windrows with KIR-1,5. In this case, the soil is enriched with organic matter. In the forest-steppe after harvesting the predecessor do stubble cultivation (BIG-3, LDG-15) to kill weeds, create a mulch layer of soil, and late autumn do the main flat tillage on medium loamy soils of the forest-steppe Priobie to a depth of 14-16 cm, in the foothills - 20-25 cm, using KPSh-9, KPSh-5, KPG-2-150, KPG-2,2, PG-3-5 and others.

Pre-sowing cultivation should maximally conserve moisture accumulated over the winter, destroy weeds and well divide the field for wheat sowing. For this purpose, early spring harrowing is carried out - BIG-3, BMSh-15 harrows are used on stubble backgrounds, and tine harrows are used in 2 traces on the mouldboard. Pre-sowing cultivation on fields mulched with straw or with thick stubble, disc implements LDG-10, LDG-15 with subsequent harrowing BIG-3, BMSh-15, on heavy soils - erosion control cultivators such as KPE-3,8, after mouldboard cultivation and fallow fields - cultivators KPS-4 with harrowing. Cultivation is better to be carried out on the day of sowing and at the sowing depth.

Application of combined tillage units allows to perform in one pass the whole pre-sowing tillage with creation of compacted seed bed, formation of mulch layer, weed removal up to 97%, which is equivalent to chemical weeding. At the same time, due to combining several operations in one pass, increasing the working width, fuel consumption is reduced by 2 times, labour force is saved, mechanical impact on the soil is reduced. The same machines are used for winter tillage with stubble retention to a depth of up to 16 cm.

Seed tillage systems combine seedbed preparation with sowing, fertilising and rolling.

Sowing dates. Scientific practice has determined the optimal sowing dates for specific conditions. When determining the sowing date of spring wheat, it is necessary to be guided by the fact that seeds germinate at a temperature of $+5^0$C, and the grain should mature at an average daily air temperature of at least $+20^0$C and reach waxy ripeness before the onset of early autumn frosts. In more moisture-sufficient eastern and foothill zones of Altai Krai with more even distribution of precipitation, but with shorter frost-free period wheat sowing should be carried out in the first decade of May when soil warms up to $+5^0$C. When sowing in the third decade of May, yield and quality are significantly reduced due to the fact that the plants use winter moisture less effectively and are more affected by diseases, in particular rust. In late sown crops, rust damage coincides with grain filling, resulting in stubby grain. Late sown crops in the foothills are more affected by the Swedish fly, as tillering coincides with the mass summer of the flies.

Bulking and ripening at late dates is at lower temperature and high air humidity, which lengthens this period by 10-12 days and reduces grain quality (1000 seeds weight, nutrition, gluten content).
In the steppe there are few precipitations and they fall unevenly (May-June are dry, July-August are more humid). It is possible to combine the critical period for moisture in spring wheat when sowing at the end of II-III decade of May. According to long-term experiments at the Kulunda Agricultural Experimental Station, wheat yield increase at sowing in the III decade of May is 0.4 tonnes/ha compared to sowing in the I decade.
In the forest-steppe of the Altai Priobie region with good moisture reserves in the metre layer of soil of 150 mm and more, early sowing dates are optimal, so fallow fields should be sown first of all in the first decade of May. On fields with less moisture reserve, weedy with oats, it is better to sow later, i.e. 15-25 May, having previously destroyed germinated weeds.
Seed preparation. Plants produced from large seeds develop a stronger root system, grow faster, are less affected by drought, are much less affected by diseases and give a higher yield.
For sowing it is necessary to use conditioned seeds with the best sowing qualities. The weight of 1000 seeds should be 35-40 g for soft wheat, not less than 40 g for durum wheat, growth force of soft wheat - not less than 80%, durum wheat - 70%.
Air-thermal heating of seeds in the sun for 3-5 days or in dryers for 2-3 hours at a temperature of +30^{0}C increases germination and germination energy and is especially necessary for seeds ripened under heat deficiency.
Prior to sowing, seeds are treated against diseases (buntings, root rots) using dressing agents according to
"List of pesticides and pesticides allowed for use on the territory of the Russian Federation, e.g. Vitavax 75% s.p. - 2.5-3 kg/t, Maxim - 1.5 l/t, Dividend Star - KE 1 l/t. Dressing is carried out with the addition of 10 litres of water. Until recently, dressing with addition of adhesives (sulphite-alcoholic bard concentrate, technical casein, etc.) was often recommended, but nowadays dressing preparations, which already contain adhesives, are offered for use.
Dressing is done not later than 15 days before sowing, using dressing machines PS-10, APS-4.
Some experience has been accumulated in the use of humic fertilisers, growth stimulators for seed treatment and foliar treatment of plants. For example, according to the data of the Agrochemistry Department of

ASAU, preparations of SILK, sodium humate, Tellura-Bio give yield increase in wheat with seed treatment up to 1.5 c/ha, and from seed treatment + spraying in tillering phase - up to 2-3 c/ha. These preparations fulfil both protective function and are stimulators of root growth, increase field germination, relieve stress of plants when treated with pesticides. In this case, small doses (SILC: 10-20 g/ha, 50-100 g + 10 l water/t seeds or humic fertilisers: 50100 g + 300 litres of water/ha of crops) make their application effective and economically viable.

Seed sowing rates vary depending on sowing date, available moisture and soil nutrient availability, weather conditions and sowing method. Optimum seeding rates are determined for each zone. The level of yield largely depends on the density of plants by harvesting. In the steppe zone of Altai Krai it should be 200-220 pcs/m^2, forest-steppe - 300-350, foothill and eastern - 400-420 pcs/m^2 respectively. Taking into account the survival rate of plants, sowing coefficients should be as follows (Table 10).

Table 10

Optimum seeding rates in Altai Krai

Cultivation area	Seeding rate, mln germinated seeds per 1 ha
West Kulunda steppe	2-3
Eastern Kulunda, Rubtsov-Aleyskaya steppe	3-4
Priobskaya forest-steppe	4-5
Forest-steppe of the Salair and Altai foothills	5-6

In the steppe, the sowing rate is reduced so that the plants suffer less from moisture deficiency, and the plants grow more slowly and utilise summer precipitation more efficiently in a sparse state. In more humid areas with higher densities, plants use moisture more efficiently, develop faster and suffer less from heat deficit and frost at maturity.

Sowing methods. In production practice, the most common is the conventional row method (with row spacing of 15 cm, seeders SZ-3.6, SZP-3.6). Narrow-row (with row spacing of 7.5 cm, SZU-3.6), cross-row method, carried out in two passes of the seeder with half seeding rate, subsoil-spreading (SZS-2.1 with splitters in coulters, soil-processing seeding complexes PPK-8.2, PPK-12.4, Ob-4, SKP-2.1) have some advantage, as they allow to distribute seeds more evenly over the area, due to which the cultivated plants develop better, oppress each other less, increase productive bushiness and power of root system, use light, moisture, nutrients more fully, fight weeds better in the rows and give yield increase of 0.2-0.3 tonnes/ha. In narrow-row and cross-row

methods, the seeding rate is increased by 10-12%.
In arid areas where low seeding rates are used, sowing on a stubble background, the narrow-row method has no advantage over the row method. In the steppe, late sowing dates are often practised, with the topsoil drying out to 10 cm. Then shallow beard sowing is an effective method. The coulter of SZS-2,1 seeder is deepened by 10 cm, the seeds are placed in the wet layer with 5 cm soil layer above it, as the seeder works without loops, the furrow is not completely embedded. The soil after sowing with these drills remains corrugated with plant residues and is not blown out. The tillering node is formed deeper, which increases the drought resistance of the plants.
Seed sowing depth depends on zonal soil and climatic conditions and ranges from 2.5 to 8 cm. On heavy cohesive soils in areas of sufficient moisture, seeds are sown at a depth of 3-4 cm, in arid areas - at a depth of 5-8 cm. Deeper sowing provides more moisture, but leads to thinning of the stalks.
Crop care includes rolling, weed, disease, pest and lodging control.
When sowing seeds in insufficiently moist or loose unsown soil, it is useful to roll the seeds with ring rollers. It promotes closer contact of seeds with the soil, transfer of moisture from the lower to the upper layers, which contributes to the rapid and friendly emergence of sprouts and good tillering.
Harrow spring wheat sowing on heavy soils, on which a crust may form, hindering the emergence of sprouts, before sprouting or after sprouting. Pre-emergence harrowing is carried out 3-4 days after sowing, it is necessary to adjust the depth of loosening so as not to damage seedlings, wheat sprout should be no more than 1 cm, and weeds - in the phase of white thread. Both toothed and needle harrows (BZSS-1.0, BIG-3A) are used. Harrowing of seedlings is carried out when the plants are well rooted in the phase of 3-5 true leaves, for which tooth harrows or rotary hoes are used. Harrowing should be carried out across the crop.
Spring wheat is characterised by relatively slow development during the initial vegetation period. In this regard, it is necessary to carry out weed control, starting as early as possible and finishing before the wheat emerges into the tube.
Losses caused by weeds range from 7 to 50 per cent depending on the degree of weediness of the fields. Biological weed control measures are used to suppress weeds in wheat grass. This is achieved through earlier sowing dates, higher field germination of seeds, application of diagonal-cross or narrow-row sowing method, which contributes to obtaining friendly and uniform sprouts.
Control of diseases, pests, weeds is carried out taking into account the

economic threshold of harmfulness - this is such a number of pests or degree of damage to plants, prevention of which will economically justify the use of active measures of plant protection. To improve grain quality, foliar feeding with nitrogen fertilisers is carried out. The feasibility, timing and rates of foliar fertilisation are determined on the basis of leaf diagnostics (see previous subparagraph).

Studies show that on 10-15% of areas in the steppe zone of Western Siberia strong grain is formed, but less than 1% of it is harvested, because farms do not form marketable batches, mixing high-quality with low-quality batches. In order to prevent this, it is necessary to analyse grain quality (gluten content, %) 1-2 days before threshing in the fields where strong wheat is supposed to be produced, to determine the order of work with grain in the current to prevent possible mixing. The main determination of quality is made after grain processing at the current, after which the results are reported to the bread-receiving station, the order of grain delivery is agreed upon, and when receiving grain at the processing plant the control determination of quality is made. This will allow more profitable sale of high-quality grain.

Harvesting. Depending on the condition of crops and weather conditions, wheat is harvested by separate (two-phase) method or direct harvesting. Direct harvesting is carried out at full grain ripeness, when grain moisture is 18% and consistency is firm. This method is used for low-growing, thinned and over-ripened breads, short-stemmed varieties resistant to lodging, as well as in areas with high humidity during harvesting. The cutting height is set at 10-20 cm; for low-grown and lodged varieties - not more than 10 cm, for long-stalked varieties - 15-20 cm.

Separate harvesting is started in the period of waxy ripeness, when grain moisture is 35%, grain has a waxy appearance, can be cut with a fingernail, its content is not squeezed out, and it does not roll into a ball. Plants are mown into swaths by cutterbars ZhVN-6A, ZhVR-10, ZhNS-6-12. Mowing height is 15-25 cm. Swath width is 1.6-1.7 m, thickness is 0.25 m - in southern areas, 0.15-0.18 m - in moistened areas. After 3-6 days start threshing of grain with moisture starting from 20%. Two-phase harvesting is used for high-stemmed, unevenly maturing and lodging and shattering varieties, as well as on weedy crops. Stem density should be at least 250-300 plants per 1 m^2. The plants should be mown across the rows to ensure better stacking of the stems in the swaths. To pick up swaths use combine harvesters of conventional or rotary type: "Don-1200", "Don-1500", "Yenisei-1200", SK-5M "Niva", SK-6 "Kolos", SK-10 "Rotor" equipped with pickers.

The two-phase method makes it possible to start harvesting earlier, prevent losses from shattering and get dry grain suitable for delivery to the elevator without additional processing, which reduces the amount of work on cleaning and drying of grain. Two-phase harvesting is especially important in areas with a long period of bread ripening and short harvesting period. The two-phase method is also preferable for harvesting seed grain, as the seeds are less traumatised, which increases germination.

Despite the significant advantages of two-phase harvesting, it should be rationally combined with single-phase harvesting. For example, in inclement weather during the harvesting period, direct harvesting is preferable, because in these conditions the ears on the root dry out faster than in swaths.

Harvesting peculiarities of durum wheat. Durum wheat has a denser ear, so it is worse threshed, and more large glassy grain with a large convex germ is more traumatised, so to prevent losses from unthreshing and traumatisation of grain and germ, it is better to harvest this crop using double drum harvesters.

Harvesting should be done in a short time - within 6-7 days and without losses. If harvesting is delayed, grain losses increase and quality deteriorates. The quality of harvesting is assessed by the following indicators: losses behind the reaper - no more than 1%, lodged - 2.5, behind the picker - 0.5, unthreshed and unshaken - 1, crushing of food grain - 1, fodder grain - 2%. The load of spike crop on one combine harvester is about 100 hectares, in some cases it increases up to 200 hectares.

4. *FORAGE CROPS*

4.1. Importance, distribution, yield

Forage crops include barley and oats. Grain from these crops is used for both food and fodder purposes.

The glassy grain of barley is used to make pearl and barley groats. Oat protein contains all essential amino acids, the grain is rich in B vitamins, iron, calcium, phosphorus, so oat grain is used to make dietary products (oat groats, hercules, tolkono, biscuits, coffee). Flour from barley and oat grain can be added to wheat flour (15-20%) if necessary, as in its pure form it is not suitable for baking. The chemical composition of grain is shown in Table 11.

Table 11

Grain chemical composition, % of ASV

Culture	Protein	BEV	Fats	Fibre	Ash
Oats	11,7	68,5	6,0	11,5	3,4
Barley	12	64,6	2,1	5,5	2,8

Barley and oats are the main grain-forage crops. Grain, as a concentrated fodder, contains in 1 kg - 1.27 k.u. of barley and 1 k.u. of oats and is used for fattening animals. -Oats and is used for fattening animals.

Barley grain is used in the brewing industry to make malt. Double-row barley with a 1000-seed weight of at least 42 g, a natura of at least 610 g/l, equalised, low-film, with a filminess of no more than 8-10% is suitable for this purpose. Since malt is prepared from germinated barley, it is important that the germination energy is at least 90-95%. The grain should have a low content of predominantly high molecular weight protein of 9-11%, a starch content of at least 60%, and an extractivity (the content of organic matter that can be transferred into aqueous solution under the influence of barley malt enzymes) of at least 78%.

Agrotechnical importance. Oats almost do not suffer from root rots, wheat after oats is less infected by root rots, therefore oats are considered a "sanitary" crop and in farms with high agrotechnics can be a good precursor for wheat.

The area sown in the world is about 79 million ha of barley and 26 million ha of oats. In Russia, about 16 million ha of barley and 8.3 million ha of oats are sown. Barley is more widespread in the North Caucasus, the Central Caucasus, the Non-Chernozem Region, the Urals, and Siberia, while oats are more common in the Non-Chernozem Region and Siberia. Among early spring crops, barley and oats, as the most plastic crops, give more stable yields, with a national average of about 1.6 tonnes/ha, while potential yields are up to 3-7 tonnes/ha. In Altai Krai the yield is 1.5-1.6 tonnes/ha on average in the Krai, 3.0-3.5 tonnes/ha in the variety plots.

4.2. Barley classification

The genus Hordeum L. has many species, of which the species Hordeum sativum Jessen is cultivated. This species is divided into subspecies: 1) Hordeum vulgare L. - is multi-row barley or common barley, in which all three spikelets develop and produce grain in the inflorescence of the spikelet on each apex of the spikelet stem; 2) Hordeum distichon L. - is a two-row barley in which one of the three spikelets on the awn of the spikelet stem is developed - the middle one, and the two lateral ones are sterile. The fruiting spikelets are arranged in one plane, which gives the ear a double row shape; 3) Hordeum intermedium Vav. et.Orl. - barley
intermediate, one to three grains may develop on a ledge.

The subspecies of multi-row barley are divided into groups: 1) correctly six-rowed (hexagonal) barley - hexastichum L. - the ear is dense, thick, short, in the cross section has the appearance of a regular hexagon; 2) incorrectly six-rowed (tetrahedral) barley - tetrastichum L. - the ear is less dense, the rows of grains are not quite correctly arranged, the lateral spikelets go behind each other, in the cross section forms a quadrangle (Fig. 7).

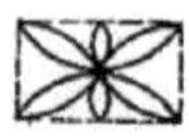

12

Figure 7. Diagram of grain arrangement in multi-row barley:
1 - correctly six-row; 2 - incorrectly six-row

The subspecies two-row barley is divided into groups: 1) nutantia R. Red., in which the lateral sterile spikelets have spikelet and flowering scales; 2) deficientia R. Red. in which the lateral spikelets consist only of spikelet scales. In crops, two-row barley is represented by the nutantia group, and the deficientia group occurs as an admixture.

Multi-row and double-row barley are widespread in our country. Multi-row barley is more rapid and drought-resistant than double-row barley, but its grain is not suitable for brewing, as it is uneven, the lateral grains are smaller and curved at the base.

Barley subspecies are divided into varieties. The most widespread in the country are varieties Nutans, Cu - Cum, Pallidum (Table 12).

Table 12 Distinguishing features of barley varieties

Variety	Spike colouring	Ear density	Jagged points	Grain foaming
Multi-row barley. Tetrastichum group				
Pallidum	Yellow	Loose	Jagged.	Plench.
Nigrum	Black	Loose	Jagged.	Plench.
Ricotenze	Yellow	Loose	Smooth	Plench.

Celeste	Yellow	Loose	Jagged.	Naked
Trifurcatum	Yellow	Loose	Instead of spicules, three-lobed appendages	Naked
Double-row barley. Nutantia group				
Nutans	Yellow	Loose	Jagged along the entire length	Plench.
Nigrikans	Black	Loose	Same	Plench.
Medicum	Yellow	Loose	Smooth, slightly serrate at the top.	Plench.
Persicum	Black	Loose	Same	Plench.
Nudum	Yellow	Loose	Same	Naked

The barley grain is broad, compressed from the sides, filmy, and the flower scales are tightly fused with the grain. Plenchability is 9-11% in double-row barley and 10-13% in multi-row barley.

The weight of 1000 seeds is 40-60 g. The grain of holo-grain varieties is richer in protein, but the film forms are used in production, as they do not crumble and are more productive.

4.3. Classification of oats

The oat genus Avena L. belongs to the bluegrass family Poaceae. Avena L. includes both cultivated and wild species of oats (oats).

The oats cultivated in our country belong mainly to the species Avena sativa L., which has 2n = 42 chromosomes.

This species is divided into two groups according to the shape of panicles: 1) dif- fusa - the panicle is spreading, with branches coming off the main axis in different directions; 2) orientalis - the panicle is single- broomed, the branches are short, tightly pressed to the main stem and in most cases directed in one direction. Single-needle forms are later maturing, more resistant to shattering, dusty bunt, and ripen more evenly (Fig. 8).

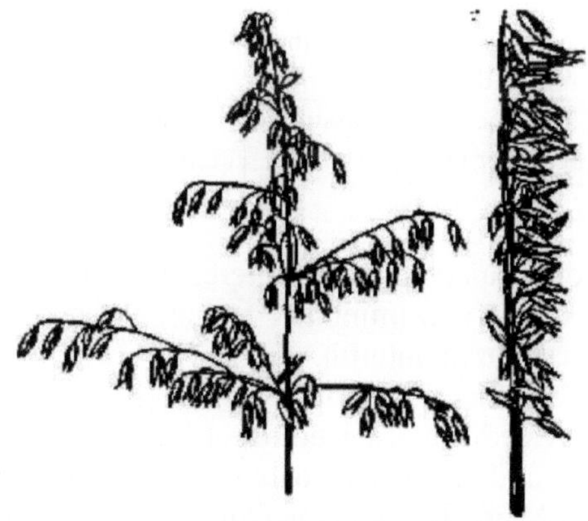

1 2

Figure 8. Main types of panicles of sown oats:

1 - spreading; 2 - single-horned

The cultivated species of oats - Byzantine and sandy oats - are of less production importance (Table 13).

Distinguishing features of oat species

Table 13

Species name	Character of the base of the lower grain	Structure of the apex of the outer flower scales	Number and nature of awns in the spikelet
Common oats (Avena fatua L.)	Horseshoeing of all grains. When falling out of the spikelet scales, all grains fall out singly	No spicules.	At least two coarse knuckles
Southern oats (Avena ludoviciana Dur.)	Horseshoeing only one lower grain. When falling out of the spikelet scales, all grains fall out together - in twos or threes	No spicules.	Two coarse patellar awns
Mediterranean oats (Avena sterilis L.)	The horseshoe is present only at the bottom grain of each spikelet. When all grains fall out together - in twos or threes	No spicules.	Two coarse patellar awns
Sown oats (Avena sativa L.)	No horseshoe. The base of the lower grain is straight, during threshing the upper grain is separated at the top of the stem	No spicules.	One awn or awnless
Oats Byzantine oats (Avena byzantina C. Koch.)	No horseshoe. The base of the lower grain is slightly bevelled, the stem of the second grain breaks in the middle during threshing	Without osteoid acuminations	Usually two awns, rarely one or absent
Sandy oats (Avena strigosa Sehr eb.)	The horseshoe is missing. The lower grain sits on the stem	With two spicules 3-6 mm long	There are usually two rough spines

Byzantine or Mediterranean oats are interesting for breeders because of their drought resistance and immunity to fungal diseases. In areas with mild climates, it is sown in autumn.

Other oat species are found in crops exclusively as weed killers. These include: from cultivated species sand oats (Avena strigosa Schreb.), also called bristly oats, and from wild oats common oats (Avena fatua L.) and southern oats (Avena ludoviciana Dur.).

White-grain varieties are more widely used (Table 14); they have larger grains but coarser straw. Yellow-grain varieties have smaller grains, but they have a lower percentage of scales, are richer in fat and vitamins, and are more drought-resistant. Grey-grain wintering forms are used in southern Europe. Brown-grained forms are used in drained bogs.

Table 14 Varieties of sown oats

Grain foaming	Colouring of flower scales (grains)	Spikelet awns	Broom shape	
			sprawling	single
Captive	White	Goatless	Mutika	Obtuzata
-	-	spinous	Aristata	Tartarica
-	Yellow	Goatless	Aurea	Flava
-	-	spinous	Krausei	Ligulata
-	Grey	Goatless	Grisea	Borealis
-	-	spinous	Cinerea	Armata
-	Brown	Goatless	Brunnea	Tristis
-	-	spinous	Montana	Pugnax
Naked	White	Goatless	Inermis	Gymnocarpa
-	-	spinous	Hinenzis	-

4.4. Biological characteristics of barley and oats

Spring barley and oats belong to early spring crops, they are long-day plants with low heat requirements. The minimum germination temperature is +1...+20C, to obtain viable seedlings it is necessary +4...+5, the optimum temperature is +20C.

+15^{0}C. At germination, barley seeds absorb up to 50% of moisture from seed weight, oats - 60%. Small frosts -4 ... -50C seedlings bear without damage, in the phase of three leaves up to -8, but in the phase of flowering and filling frosts -1,5 ... -2^{0}C lead to the death of pollen and loss of seed germination. The optimum temperature from sprouting to earing is +18...+220C, at grain ripening +23...+240C. Barley is more resistant to high temperatures. Paralysis of stomata at temperature +40^{0}C occurs in barley in 25-30 hours, in oats - in 4-5 hours, in wheat - in 10-17 hours. The sum of active temperatures for full development is 1000-1500^{0}C for early maturing and 1800-2000^{0}C for late maturing varieties. These crops are characterised by self-pollination, with barley being a strict self-pollinator, flowering even before the ear emerges.

Among early spring crops, barley is the most drought-resistant, salt-tolerant, and sparingly consumes moisture. Transpiration coefficient of barley - 350-400.

The increased drought tolerance of this crop is associated with its early maturity and ability to utilise nutrients intensively during early growth phases. Especially drought tolerant are the early maturing multi-row barley crops.

Oats are more water-loving than barley and wheat, but they tolerate early spring droughts better due to their rapidly developing root system. The transpiration coefficient of oats is 474.

The critical period in terms of moisture is tube emergence - earing (sprouting in oats), when the ear and panicle are formed in embryonic state. In oats, the critical period is longer and includes tube emergence, spikelet emergence and flowering. If there is a lack of moisture during this period, the number of infertile spikelets increases. Barley is resistant to "fires" and "seizures", but early spring drought is worse tolerated than oats because of the weak development of the root system during this period.

The period of intensive consumption of nutrients in barley is short, which, along with the low digestive capacity of the roots, makes barley more demanding of soil fertility. Medium-bound loamy fertile soils are suitable for barley; acidic, waterlogged, light sandy soils without appropriate improvement are of little use. Optimal pH 6,8-7,5. Barley grows better on slightly saline soils than wheat and oats.

Oats are less demanding on soil fertility than wheat and barley, as their root system has the ability to extract hard-to-solubilise nutrients from the soil. It can grow on loamy, clayey, peaty soils. It succeeds better than other breads on acidic soils with pH 5-6. Solonetz soils are not suitable for it.

Good malting qualities of barley are formed when it is grown in areas with a moderate temperature regime and sufficient moisture, with relatively low-humus soils with good soil fertility and good soil quality.

soils with good

water permeability.

4.5. Cultivation technology

Barley and oats in crop rotation are placed after row crops (maize, potatoes, beetroot), leguminous crops, winter crops, fallow crops, perennial grasses. It is not recommended to place oats after beetroot, as they have a common pest - nematode. When placing barley in the crop rotation it is necessary to take into account the purpose of cultivation. After legumes, perennial grasses barley grain contains a lot of protein, which is important when using it for fodder and food purposes. Tillage leaves the fields clean of weeds, significantly depleting the soil of nitrogen, so malting barley sown on these predecessors gives a high yield of grain with a high starch content. It is not recommended to sow malting barley on fields with high fertility and rich in nitrogen.

Autumn (main) tillage. In steppe areas with soils of light mechanical composition early (immediately after harvesting the predecessor) shallow tillage on 10-12 cm with KPSh-5, KPSh-9, OPT-3-5 implements with leaving 80% of stubble on the surface to protect the soil from erosion is effective. In more humid areas of forest-steppe and foothills, after harvesting the predecessor, stubble surface treatment is made with huskers, needle harrows or disc harrows to kill weeds, create

a mulch layer and prevent moisture evaporation. After
two weeks after surface tillage in late autumn, the main flat tillage is carried out: on
medium loamy soils of Priobskaya and Biysko-Chumyshskaya zones of Altai Krai by 14-16 cm, on sloping lands - by 2022 or 25-27 cm for better absorption of melt water and reducing its runoff. On heavy soils in the foothill zone a good result is given by cultivation with C - BIME stands, cheesel ploughs at 27-30 cm across the slopes. Ploughing is acceptable after silage row crops (maize, sunflower) and perennial grasses.

Early spring tillage begins with early spring harrowing at 4-5 cm for mulching and levelling of soil with BIG-3A, BMSh-15 on flat-cutting tillage, and on moulded grain - with tooth harrows across the main tillage. The clumpy surface is rolled.

Pre-sowing cultivation in the steppe zone on clean fields is combined with sowing when using seeder-cultivator SZS-2,1, seeding complexes of "Kuzbass" type. Clogged fields on the day of sowing are cultivated to the sowing depth by erosion control cultivators with rolling. In the forest-steppe zone on mouldboard and sparse stubble no-tillage backgrounds pre-sowing treatment is done with KPS-4 cultivators with harrowing, in the presence of dense stubble - with harrowers with subsequent harrowing with needle harrows or KPE-3,8 with harrowing and rolling.

Consumption of up to 50% of nutrients in barley occurs in the tillering - earing phase, and the root system develops quite slowly, so a high supply of nutrients is necessary. Nutrient inputs are shown in Table 15.

Table 15

Removal of nutrient elements per 1 tonne of grain and related by-products, kg

Culture	N	P	K
Barley	25-30	10-12	20-24
Oats	27-31	10-12	22-29

Oats, due to their well-developed root system with high digestive capacity, effectively utilise soil fertility and nutrients after the predecessor and after-effects of organic matter. It is better to apply organic fertilisers under the predecessor 15-20 t/ha. When growing barley, it is recommended to apply the full rate of nitrogen fertilisers in combination with phosphorus and potassium fertilisers. When calculating the rate of fertiliser application, it is necessary to take into account nutrient removal, their content in the soil, coefficients of use from the soil and fertiliser.

For malting purposes, the rate of nitrogen fertilisers for barley is reduced by 20-25%. In the experiments of ANIIZiS application of fertilisers in growing malting barley $N_{45}P_{60+20}K_{60}$ provided an increase of 0.37-0.59 t/ha of grain, and in humid years - up to 1.3 t/ha.

Preparation of seeds for sowing. For sowing it is necessary to use conditioned seeds with purity not less than 97%, germination not less than 87% (GOST R 52325-2005), with a weight of 1000 seeds 38-40 g. Oats form 2-3 grains in a spike, the lower grains are larger and give a higher yield, so it is advisable to select them for sowing on triers. Before sowing, seeds are subjected to air-heat heating, which increases their germination by 10%. Seed dressing for destruction of seed and soil infections by systemic preparations Dividend Star KE 1l/t, Premis 1.5l/t and others. During dressing it is advisable to use microfertilisers: magnesium sulphate (0.7 kg/t), zinc sulphate (0.8 kg/t), sodium humate (0.75 kg/t), which increases yield by 0.25 t/ha.

Sowing. Agrotechnical terms of sowing early spring crops of barley and oats come at a soil temperature of +5 ^{0}C at the sowing depth. In Altai Krai soil warming up to such temperature coincides with physical ripeness of soil in early May. In more humid forest-steppe areas of the foothills with a relatively even distribution of precipitation during the growing season, these cold-resistant crops should be sown in early May. In this case, the plants more effectively use the good reserves of winter-spring moisture, and, in addition, the early sowings have time to finish the tillering phase by the time of mass summer of the Swedish fly and are less damaged by this pest. It is also better to sow malting barley as early as possible because the grain is formed with higher germination energy.

When growing grain forage in arid conditions of steppe zone with uneven distribution of precipitation, when moisture reserves before sowing in the metre layer of soil do not exceed 150 mm, it is better to sow them in later dates of 15-25 May so that the critical period for moisture output in the tube - earing (spikelet) coincided with a large amount of precipitation in July, and the early spring period to use for provocation and destruction of oats.

Seeding rate of barley and oats by zones of the region is differentiated as follows: in Kulunda steppe - 2-3 million germinated seeds per 1 ha, in Aleyskaya steppe - 3-4, in forest-steppe of Priobie - 4-5, in forest-steppe of foothills - 5-6 million pcs/ha. Malting barley is grown in dense crops (seeding rate - 5.5-6 million pieces/ha), so that there was less podding (side stalks with small ears), which means that the grain will be more uniform. In addition, in such crops barley matures more quickly, the protein content in the grain decreases, which means that the starch content increases.

The optimum sowing depth is 4-6 cm, with the seeds lying on a moist compacted layer of soil. On heavy soils, the depth should be reduced to 3-4 cm and on light soils increased to 6-8 cm. Oat seeds can be sown slightly deeper, as it has a longer coleoptile and practically does not react to sowing depths of 3-8 cm.

Barley and oats are sown most often by row sowing with row spacing of 15 cm. At late sowing dates in the steppe, when the topsoil is drying out, the shallow lateral sowing method, when the seeds are at a depth of 10 cm with 5 cm of loose soil above them, gives good results. For this purpose the seeder SZS-2,1 is used. In more humid conditions the cross or narrow-row sowing method is advantageous, where the yield can be higher by 0.2 tonnes/ha due to more even distribution of plants and more efficient use of light, water, nutrients, reduction of weeds. Malting barley sown with

narrow-row or cross-row crops produce more uniform grain in terms of grain size and ripeness. Yield and quality increase when rows are orientated from south to north.

Sowing care. After sowing it is necessary to roll the seeds to equalise the depth, improve the thermal and water regime of the soil, increase field germination, but if the soil is wet at sowing, it is better to sow without rollers, and roll the next day with ring rollers. Weed control is done by means of pre-emergence harrowing 4 days after sowing, when weeds are in the phase of white thread, post-emergence harrowing, when barley plants are in the phase of 3 leaves - tillering. During the vegetation period, control diseases, weeds and pests taking into account economic thresholds of harmfulness with the use of preparations authorised for use.

Harvesting. Barley ripens in a timely manner. At the onset of full ripeness the ears become brittle and droop, so if harvesting is delayed, part of the crop is lost. Barley is harvested separately at the waxy ripeness phase when grain moisture is 35%. Direct harvesting starts at full ripeness at a moisture content of no more than 20%. Malting barley should be harvested in the phase of full ripeness directly, because in the plants on the root from waxy to full ripeness in the grain is established a favourable combination of protein and carbohydrates. Threshing should be carried out at a moisture content of 15-16% at 1000-1400 revolutions of the drum per minute, so that the grain is less traumatised. Grain is also less traumatised when threshing in double drum harvesters. If the fields are clogged and barley does not ripen at the same time, it is more appropriate to use separate harvesting.

Prior to harvesting, batches of grain are formed homogeneous in quality: valuable fodder grain, brewing grain. For this purpose, grain samples

are taken from the fields for preliminary assessment of protein content, coarseness, uniformity and other indicators. Upon arrival at the current, the seeds should be cleaned, dried and sorted. The drying regime for malting barley is the same as for seed material, i.e. it is not allowed to heat the seeds to a temperature of more than 45^0C.
Oats have a longer ripening time than barley and wheat because they have a panicle inflorescence, the upper part of which ripens faster. The difference in grain moisture content can be up to 5%, at the top the grain is ripe and has a moisture content of 17, at the bottom the grain has a moisture content of 23%. In addition, the straw in the lower part of the oats remains green and moist for a long time. Therefore, oats are more often harvested separately, starting mowing when the grain in the middle part of the panicle is waxy ripe. It can also be harvested directly at the beginning of full ripeness.
In Siberia, feeding animals with whole plants of grain-forage crops in technologically processed form (silage, haylage, dry monofeed in the form of pellets or briquettes) is of interest. Full threshing-free harvesting of barley, oats or their mixtures starts at the phase of milk-wax ripeness, plants are mown, chopped, used for haylage, silage or dry monofeed.

5. *CUCURUZA*

5.1. Importance and utilisation, distribution and yield

Maize is one of the most productive and widespread crops. 25% of the grain produced is used for food, 60% for livestock feed and the rest for processing. Corn grain contains from 9 to 12% protein, 77% starch, 4-8 fat, 2 ash, 2.5% fibre, mineral salts and vitamins. The grain is used to produce flour, groats, cereals, flakes, preserves, starch, alcohol, beer, glucose. Oil and vitamin E are extracted from the germ. Stigmas are used in medicine. In the USA and other countries, 50% of sugar needs are met by the production of molasses, starch sugar, glucose from corn.
Grain is used for livestock feed. 1 kg of grain contains 1.3 k.u., but not enough protein. There are 78 g of digestible protein per 1 k.u. in maize grain, whereas according to zootechnical norms 115 g are required.
Maize in Siberia is, first of all, a silage crop. In 100 kg of silage from maize with cobs of waxy ripeness contains 21 k.u. and 1800 g of digestible protein. Silage has a good digestibility and has molokogonichnymi properties. Maize is also used for green fodder, rich in carotene, but the feed value of green mass without cobs is 2 times less.
Maize is a row crop and is a good forerunner, leaving fields free of weeds.
Maize ranks first in the world in terms of gross yield and second after wheat in terms of area under crops, covering 132 million ha (USA, Argentina, Brazil, Mexico, China, India, Europe) with an average yield of 3.7 tonnes/ha. In Russia, the area sown for grain cultivation is 0.5 million ha (Stavropol, Krasnodar Territory, Volga Region, Central Central Central Zone), with an average yield of 1.7 tonnes per hectare. In the USA, grain yield is 6 tonnes per hectare. For silage, this crop is grown everywhere, including the Black Earth Region, Siberia, and the Far East, with a yield of about 20 tonnes/ha, although the potential yield is 30-50 tonnes/ha, and up to 80 tonnes/ha under irrigation.
In Altai Krai, maize for grain is sown within 1500 ha in the steppe Kulunda zone with yields from 0.5 to 2.3 tonnes/ha in different years. The area of maize sown for silage in Altai Krai has significantly decreased to 300 thousand ha, yield - 17 tonnes/ha.

5.2. Classification and botanical characterisation maize

Maize (Zea mays L.) belongs to the bluegrass family (Poaceae). The species of cultivated maize include subspecies that differ in traits (Table 16).

Table 16 Distinctive grain characteristics of different maize subspecies

Subspecies	Grain surface	Top of grain	Horn-shaped endosperm	Mealy endosperm	Grain (size and shape)
Tooth-shaped (inden- tata).	Smooth	Notched	Develops only on the sides of the grain	Developed in the centre of the grain and at the apex under the indentation	Large, elongated
Siliceous (indù- rata)	Same	Rounded	Strongly developed, almost solidly performing grain	Available only in the centre of the grain	Small, rounded, compressed on ventral and dorsal sides
Starchy lacea (amy- lacea)	Same	Same	Absent	Goes with the grain	Same, but coarse grain
Bursting (everta)	Same	Rounded or pointed	Strongly developed	Present in embryo or absent	Shallow, rounded, slightly compressed
Sugar (saccharata)	I'm a wrinkle artist	Wrinkles and wrinkles	It's all grain.	Absent	Large and medium-sized, rounded, squashed

The root system is axillary, well developed, capable of penetrating up to 3 m in depth, has 4 types of roots: 1) germinal root; 2) lateral roots on the germinal root form the primary root system; 3) secondary roots from underground stem nodes; 4) auxiliary supporting roots from lower above-ground stem nodes. The main mass of roots is in the layer of 50-60 cm. The stem is erect, rounded, parenchyma, 0.5 to 3 m high, sometimes branching in the above-ground part, forming shoots. Leaves are linear, large, from 8 to 25 on one plant. There are two types of inflorescences: male - the panicle is located at the top of the stem or side branches, and female - the cob is located in the axils of leaves. Maize is a cross-pollinating, wind-pollinated plant. Female flowers have a pistil with a large ovary and a very long column that extends beyond the cob wrapper during flowering. The male inflorescence blooms 3-8 days earlier, which ensures cross-pollination.

The fruit is a naked seed pod of white, yellow, cream, orange colour. The weight of 1000 seeds is from 100 to 499 g.

For sowing maize, hybrid seeds are used, which are obtained by crossing varieties or self-pollinated lines. Depending on the parental forms, intervarietal, varietal-linear, inter-linear hybrids.

Hybrids are 25-30% more productive than varieties due to the effect of heterosis. Hybrid seeds of the first generation provide the maximum yield increase; after reseeding, the heterosis effect is significantly

reduced, so hybrids are reproduced annually. To compare hybrids from different countries of the world in terms of maturity, the Food Organisation of the United Nations (FAO) has developed a scale of maize maturity classes. At present, when assigning a numerical number to new hybrids, it is necessary to follow the FAO maturity classification (Table 17).
Recommended ratio of sowing areas of hybrids for Western Siberia: very early maturing hybrids - 25%, early maturing -
35, medium-early - 20, medium-ripening - 20 per cent.

Table 17

Classification of hybrids by maturity groups

Ripeness group	Number of leaves	Growing season, days	Sum of active temperatures, ^{0}C	FAO ripeness group
Very early maturing	Up to 11	85	2100	100-149
Early maturing	12-14	90-10°	2200	150-199
Medium-early	15-16	105-115	2400	200-299
Mid-maturing	17-18	115-120	2600	300-399
Mid-late	19-2°	120-130	2800	400-499
Late maturing	21-23	135-140	3000	500-599
Very late maturing	Over 23	145-150	More than 3,000	Over 600

5.3. Biological characteristics of maize

Maize is an annual herbaceous plant. It is a short-day and heat-loving crop. Its homeland is Central and South America, a zone of tropics and subtropics. This is the reason for the biology of the plant.
Minimum germination temperature is +8...+10^{0}C, but in early maturing, more cold-resistant hybrids - +6...+80C. The temperature of viable seedlings +10 ... +120C, but germination is very slow (18 days), seedlings are affected by diseases, rot. At a temperature of +21... .+25◦ C seedlings appear in 7 days.
From sowing to maturity, maize requires a sum of active temperatures ranging from 2°°°° C for early maturing and up to 26°°°- 3°°°° C for late maturing hybrids. Silicea maize is more cold tolerant and early maturing than denticle maize.
A prolonged frost of -3° C damages maize seedlings, but when the leaves die off, the plants remain alive, because until the formation of 6-7 leaves, the growing point from which the plant develops remains in the soil and is not damaged. After a few days, growth resumes. But such autumn frosts severely damage the plants and the seeds lose germination.
The optimum temperature for growth and development is +25.+30° C.

Maize belongs to mesophytes, but it can be referred to relatively drought-resistant crops, as it has a well-developed root system up to 3 m and is characterised by the ability to use moisture more sparingly. The transpiration coefficient is from 190 to 370, whereas in the breads of the first group it is from 300 to 600. But the total water consumption is quite high, as maize forms a large yield of dry matter. Late spring crops, to which maize belongs, unlike early spring crops, use precipitation of the second half of summer more effectively, as the differentiation of the growth cone occurs later. This in the conditions of Altai Krai provides them with drought resistance.

Due to its grooved shape and oblique-vertical leaf arrangement, maize makes good use of even minor precipitation and dew running down the stem to the roots. At the beginning of the growing season, maize is slow growing and consumes little moisture and nutrients. Water consumption increases at the beginning of the tube emergence, and the critical period for water consumption is 10 days before emergence and 20 days after flowering, as generative organs, particularly pollen, are formed at this time. Drought during this period causes pollen sterility, deterioration of fertilisation, and through the laceration of male and female inflorescences from 2-3 to 20 days. Male inflorescences bloom earlier.

Maize is a short-day light-loving plant, flowering faster on a short 8-9-hour day. When moving northwards, under conditions of a long 12-14-hour day, the vegetative period lengthens, maize does not flower for a long time, but continues to grow. Due to this, the yield of vegetative mass increases.

Maize has high requirements to soil fertility and physical properties. It grows well on loose, fertile, well water-permeable, but water-absorbent soils (chernozem, chestnut, dark grey forest, river load). Unsuitable soils are acidic (with pH below 5.0-5.5), saline, heavy with poor aeration, sandy.

The following phases are observed in maize: the beginning and full emergence of seedlings, 2nd-4th leaf (at this time differentiation of the rudimentary stem occurs, height and number of leaves are laid down), 4th-8th leaf (panicle formation), 7th-12th leaf (cob formation), panicle emergence, panicle flowering, cob flowering, lactic, lactiferous, waxy and full ripeness.

5.4. Cultivation technology

In Altai, maize is grown mainly for silage, but in the steppe zone, which is more warmly supplied with heat, it can be grown for grain if early maturing hybrids are used.

In the 80's in Siberia and Altai transition was made from cultivation of mainly late maturing hybrids, giving high yield of green mass, but low yield of fodder units, to cultivation of maize on "grain" technology for silage with cobs. In this case, predominantly early-maturing and medium-maturing hybrids are used, giving at some reduction in the total biomass yield, an increase in the amount of dry matter due to cobs of different maturity.

Place in crop rotation. Maize is placed as the second crop after fallow wheat, winter rye, after leguminous crops. In forage crop rotation - after annual grass-legume mixtures. It should not be placed after sunflower, beetroot, as they strongly dry up the soil and take out a lot of nutrients, as well as after millet, sorghum, as they have common diseases, pests and weeds with maize. On the layer of perennial grasses maize is more affected by wireworm. When applying organic and mineral fertilisers and herbicides, maize can be grown continuously for 6-8 years on a brood field near the farm, which facilitates the introduction of large doses of organic matter and reduces the cost of transporting products. Corn for grain is better located on gentle slopes of southern exposition, where the sum of active temperatures is 80-100^{0}C higher. This crop is a good precursor for spring wheat, as it frees fields from weeds and has almost no common diseases and pests with early spring wheat.

Tillage. In steppe and forest-steppe on soils subject to erosion after stubble predecessors, BIG-3A stubble tillage is made in 2-3 traces, then - no-tillage at 22-25 cm. In more humidified foothill areas, on heavier soils maize responds well to deep mouldboard ploughing . When cultivating on permanent plots, after harvesting maize in autumn, the field is disced, organic fertilisers are applied and mouldboard ploughing is done. In spring - closing of moisture, levelling. The first cultivation is done to provoke weeds, pre-sowing cultivation - at the sowing depth and on the day of sowing with rolling.

Fertiliser. Maize responds well to fertiliser.

The output per 1 kg is:

	N	P	K
for grain	2,5-3,4	1-1,3	2,5-3,7
when grown for silage	0,3	0,15	0,4

According to the data of ANIISKhoz, introductio n 20 tonnes/ha manure increases yield by 40%. The gain of maize for silage from the application of mineral fertilisers $N_{60-90}P_{60}K_{60}$ is 7.2 t/ha. With the application of 20 t/ha of manure and mineral fertiliser $N_{90}P_{60}K_{60}$ increases silage yield from 20 to 35 t/ha, protein content - by 1.2%, and protein harvest - by 2 times.

Sowing. The "grain" technology requires earlier sowing to guarantee cob production. Seeds are more susceptible to diseases in less warm

soil. In addition, maize is sown in a precise dotted pattern in small quantities, which means that as few plants and sprouts as possible must fall out. Therefore, very great attention is paid to the quality of the seed. F_1 hybrid seeds are used, although with less intensive technology it is possible to use F2 seeds or population varieties.
The choice of hybrid by zone can also be important. In the steppe zone, it is better to use early maturing hybrids, as having a faster development, they form low plants, less leafy, less moisture consumption, but due to the guaranteed formation and maturation of the cob give a good collection of dry matter per hectare. In the foothills, the frost-free period is shorter, the soil warms up slower, and sowing is later, so plants often do not have time to form a full-fledged cob. But at the same time, the total moisture availability is higher. Therefore, using medium-ripening and medium-late hybrids with a higher potential yield (in this case - vegetative mass), it is possible to ensure a high dry matter yield at the expense of a high yield of green mass when sowing at a later date.
Seeds are delivered to farms ready for sowing, calibrated, dressed or treated by inoculation, which consists of seed treatment using film-forming polymer with the addition of dressing agent, nutrients, microelements, physiologically active substances. Such seeds are better protected, have high growth vigour and can be sown earlier.
One of the reasons for low maize yields in Western Siberia is late sowing dates, which shorten the growing season, reduce the yield and its quality, as the cobs do not have time to form. On the one hand, maize is a heat-loving crop and gives friendly sprouts when sown in soil warmed up to +10...+12⁰C. According to average annual data, such temperature regime in the soil is observed, and it allows to sow corn in the steppe zone of Altai at the end of I - II decade of May, in forest-steppe - in II - beginning of III decade of May, in foothills - III decade of May. And when using early maturing, more cold-resistant hybrids, sowing with encrusted seeds is possible 5 days earlier. Positive experience was also obtained when sowing maize in pre-cut ridges, which allows to start sowing earlier, as the soil in the ridges dries and warms up faster. This is relevant in areas with a lack of heat and excessive moisture.
To protect early sowing from frosts, the sowing depth must be maintained. At frosts of -6° C the sowings with sowing depth of 6 cm perish, while at sowing depth of 8-10 cm they grow without visible consequences, even if they are damaged in the above-ground part, as the growth point remains in the soil for a long time. The optimum sowing depth in the steppe is 5-8 cm, in the forest-steppe - 4-5 cm. Sow more cold-resistant, early maturing hybrids first, then medium maturing hybrids.
Maize is sown in wide rows with row spacing of 7°, 6° or

45 cm. A more progressive method is precision, dotted sowing with precision seeders (SUPN-8, SPCH-6, SKPP-12, STV-12, etc.), as the plants are more evenly arranged in the row, receive the same optimal area of nutrition.

In dense crops the formation of generative organs slows down, fewer cobs and grains are laid in them. Therefore, "grain" technology provides for sparse crops. The recommended plant density for harvesting is given in Table 18.

Table 18 Corn plant density recommended in Altai Krai

Hybrids	Steppe	Forest-steppe
When growing for silage		
Soon-to-ripen	45-50 thousand/ha	65-70 thousand/ha
Medium-early and medium-maturing	40-45 thousand/ha	60-70 thousand/ha
When growing for grain		
Soon-to-ripen	40-50 thousand/ha	40-50 thousand/ha
Medium-early	40-45 thousand/ha	-

To ensure such density by harvesting, the calculation of seeding rate should take into account field germination (80-60%), shredding of plants during harrowing and inter-row cultivation (5% for each treatment). As a result, the weight seeding rate is 20 to 25 kg/ha.

Crop care. Rolling is carried out after sowing to ensure seed to soil contact. Maize grows slowly in the first period, so weed control is important. This is done by pre-emergence harrowing 3-5 days after sowing, when weeds are in the white thread phase.

Post-emergence harrowing is carried out at mass emergence of weeds and at 2-3 leaves formation in maize. During the growing season, inter-row cultivation is carried out using KRN-4,2. The first - in the phase of 3-4 leaves at a depth of 6-8 cm, then - when weeds appear at a depth of 4-6 cm usually after 2 weeks. This destroys up to 80% of annual weeds and improves soil aeration.

With the use of herbicides on soils of light and medium mechanical composition the need for inter-row cultivation is reduced. Modern herbicides are applied mainly during corn vegetation in the phase of 3 maize leaves, e.g. Dialen super - 2-3 l/ha, suppressing dicotyledonous weeds.

Harvesting. Maize for grain is harvested in two ways:

1) harvesting in cobs begins at grain moisture content of 40%, use a 6-row self-propelled combine KSKU-6 and a three-row trailed combine KKP-3;

2) harvesting in grain begins at grain moisture content of 30%, combine harvesters with PPK-4 attachment are used.

In both cases, the leaf mass is harvested at the same time as the grain part is harvested, chopped and used for silage.

Cleaning and drying complexes KZS-200 are used for grain drying. The cobs are dried in special chamber-type dryers or active ventilation units. These are high energy inputs.

When harvesting for grain, large losses occur during threshing and cleaning. The accumulation of dry matter and digestible protein in maize is going on from the beginning of filling to waxy ripeness, and after that until full ripeness the nutrition and digestibility of feed from plants decreases due to the accumulation of fibre. Therefore, the main method of harvesting maize for silage is mowing of plants with cobs of milky-wax or wax ripeness at total moisture content of plants 65-70%, which is optimal for silage. Plants are mown at a height of 10-12 cm using self-propelled combines KSK-100, E-280 and chopped into 5-10 cm particles. The higher the moisture content of the mass, the larger the particles should be. In order to preserve the nutritional content of the fodder, canning, i.e. ensiling, is carried out. The essence of the biological method of ensiling is that the transformation of green mass into silage occurs under the action of lactic acid bacteria, as a result of the vital activity of which lactic acid accumulates. It protects the forage from rotting and decomposition, having a destructive effect on the putrefactive bacteria. Sugar compounds are necessary for the development of lactic acid bacteria, which are sufficient in maize. It is also necessary to compact the silage mass, displace air and create anaerobic conditions, which creates better conditions for lactic acid bacteria. As the silage mass is loaded into trenches, it is compacted with heavy tractors and covered with polyethylene after filling.

CHAPTER 6

6. PROSO

6.1 Significance and dissemination

Millet is one of the most important cereal crops in our country. Millet is used to produce millet, which is one of the first among other cereals in terms of flavour and nutritional advantages. It has a high content of protein (12%) and fat (5.5%), second only to oat groats, easy digestibility and good digestibility. Wastes of millet processing into groats (flour, mullein, husk) are used to feed livestock. Millet grain in whole and milled form is widely used as concentrated fodder for poultry and fattening pigs (1kg of it contains 0.97 k.u.), as well as for making malt.

Millet can be sown for green fodder and hay, as well as used for reseeding dead winter and spring crops and as a stubble crop (after harvesting winter crops and other early maturing plants). In summer sowing millet is a good cover crop for perennial grasses.

In world agriculture, the area sown with millet is about 40 million hectares. The main producers are Russia, China and Mongolia. Millet is also sown in Japan, India, Afghanistan and Turkey. In Europe, this crop is widespread in Hungary, Poland and Bulgaria. Minor areas are found in Africa and in the eastern states of the USA.

Russia has the largest area of millet sowing in the world, ranging from 0.7 to 1.9 million hectares. About 70% of the crops are concentrated in the Volga region, the Central-Central Zone, and the Rostov region. Significant areas are concentrated in Kazakhstan, Western Siberia, the North Caucasus and Ukraine. The average millet yield in Russia is 1.19-1.55 tonnes/ha, but advanced farms produce 3.0-4.5 tonnes/ha. The area sown in Altai Krai ranges from 64.8 to 96.3 thousand hectares, with an average yield of 0.7-0.8 tonnes/ha.

6.2 Classification and botanical characterisation

Common millet (Panicum miliaceum L.) belongs to the genus Panicum. This species is characterised by the structure of the panicle, in which the lateral branches are elongated (Fig. 9). Bristle millet (Italian millet) (Setaria italica), which has a sultan inflorescence, i.e. a panicle with shortened lateral branches, is also used in culture.

shortened lateral branches. Italian millet includes two subspecies: S. italica - Italian millet proper, or chumisa, and S. italica - Italian millet. S. italica mogarium - mogar. Italian millet, or chumiza, differs from mogar by its longer spike-like panicle (1530 cm) and the presence of lobes in the panicle. It is cultivated in small areas to produce grain of food value (cereals, flour). Mogar has an implicitly lobed panicle. It is grown mainly for grain (poultry feed), hay and green fodder.

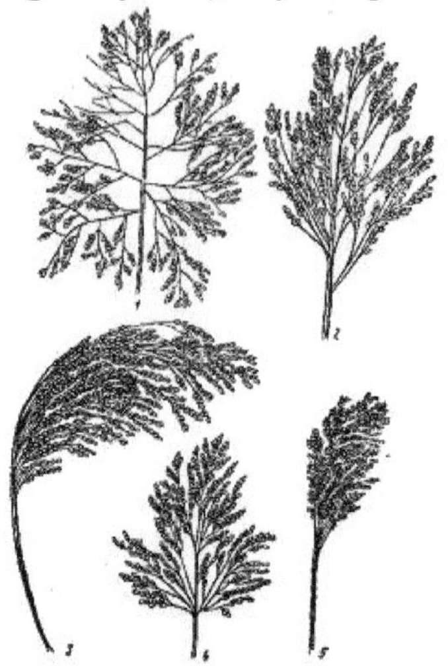

Figure 9. Panicles of subspecies of common millet:

1 - sprawling;
2 - a big one;
3 - compressed (droopy);
4 - oval (semi-comic);
5 - lumpy

Common millet is the most widespread in crops. This species is divided into 5 subspecies depending on the panicle shape, its length and density, the degree of compression and spreading of lateral branches, and the presence of pads (thickenings at the base of branches) (Table 19).

Millet straw stalk, erect, simple or branched, weakly pubescent, divided into 4-10 internodes. Stem height is from 45 to 150 cm. The number of stems in a bush forming normally developed panicles is usually 3-4, and up to 20 in case of large feeding areas. This ability of millet is utilised in broad-row and stubble sowing.

Subspecies of common millet

Table 19

Signs of a broom	Spreading millet *Subsp. patentissimum* Popov	Hanging millet *Subsp. Effusum AI*	Compressed millet *Subsp. Contraktum Al*	*Ovatum* millet (semicomplex) *Subsp. Ovatum* Popov	Fodder millet *Subsp. Compactum Korn*
Length	Long	Long	Long	Short	Short
Density	Loose	Loose	Loose	Dense	Dense
Awn	Straightforward	Straightforward	Bent	Straightforward	Straightforward
sprawl	Sprawling	Semi-spreading	Compressed	Semi-scattered	Compressed
Branch deviation	All the branches are strongly deviated from the broom axis	Lower branches deflected, upper branches pressed against the panicle axis	All the branches pressed against the broom axis	Lower branches deflected, upper branches pressed against the panicle axis	All the branches pressed against the broom axis
Pads at the base branches	All branches have	Only present on the lower branches	Absent or weak pronounced	Only present on the lower branches	Absent

The root system is coarse. It sprouts by a single root, and then secondary roots are formed from tillering nodes. The power of roots is determined not so much by the depth of occurrence up to (105 cm) as by their spread in width (penetrates up to 115 cm) and the number of root shoots (up to 120 pcs.), the bulk of roots is located in the layer 0-20 cm, and at a depth of 40 cm 80% of roots penetrate. Millet develops aerial support roots from the lowest above-ground nodes of the stem.

They increase plant resistance to lodging and drought. Millet has time to form a root system before fluttering and utilises moisture more fully than other breads. However, in dry weather the formation of nodal roots is delayed, and half-lodged seedlings live for a long time at the expense of germinal roots.

Leaves consist of leaf sheath and leaf lamina of linear-lanceolate form, pubescent or naked, green, light or dark green, in some varieties with anthocyanic colouring and clearly expressed main nerve. Leaf lamina is 18-65 cm long and 1.5-4 cm wide.

Inflorescence is a panicle 15-20 cm long, strongly branched (from 10 to 40 branches and more), one spikelet sits at the ends of the branches, in which one (upper) flower develops and bears fruit, the second (lower) flower is usually underdeveloped, has the appearance of whitish and colourless film. The flowers are oviparous, have three stamens and an ovary with two pinnate stigmas. Millet is a facultative self-pollinator, with 20% cross-pollination.

The fruit is a filmy granule, usually called a grain, of rounded, oval or elongated shape. The weight of 1000 grains is 4-9 g, with a plenum density of about 12-20%. A normally developed panicle has 600-1200 grains. The largest grain is in red-grained millet with drooping panicles.

1.3. Biological peculiarities

Millet is a heat-loving plant. Seed germination starts at a temperature of +8...+10◦ C. At a temperature of +20... .+25◦ C seeds germinate in 3 days, at +8◦ C - in 1015 days. The biologically optimal temperature at which the most vigorous germination of seeds takes place is +20...+300C, and the maximum temperature at which it is suspended is about +40◦ C. In dry weather, the formation of nodal roots is delayed, and seedlings live for a long time (15-20 days) at the expense of germinal roots. In moist soil, nodal roots grow quite rapidly and after 152° days penetrate 4°-5° cm deep.

Sprouts die at frosts of -3° C. The sum of active temperatures during the growing season is 1800-21°° C.

The heat tolerance of pearl millet (especially after hatching) is much greater than that of wheat, barley and rice. In the experiments, leaf stomata at temperatures of +38...+400C continued to function for 48 hours.

Millet is less demanding of moisture than other breads. Its seeds need only 25% of their weight in moisture to germinate. The transpiration coefficient is 200-250. The root system has a great sucking power and is able to extract moisture from the soil in case of a significant deficit.

A characteristic feature of millet is its high drought resistance. It is explained by the ability of the plant to temporarily suspend growth (during drought it seems to fall into a state of anabiosis), roll up its leaves and spread the above-ground part on the ground, which reduces evaporation of moisture.

The critical period (i.e. the highest moisture demand) is during the time of emergence to grain formation. The better the plants are supplied with moisture and nutrients at this time, the higher the yield. Millet makes very good use of the precipitation that falls in the second half of summer, when it is almost useless for breads of group I .

Millet is a light-loving plant and needs to accumulate a large amount of organic matter in a short growing season. Spreading forms of millet are less demanding of light than clumping and drooping forms. The highest intensity of photosynthesis is observed during the period from the beginning of grain filling to full ripeness, so cloudy weather in the

second half of the growing season depresses millet and significantly prolongs the growing season. Millet is a typical short-day plant. As it approaches the northern limit of its cultivation, it significantly lengthens the growing season.
Millet is not very demanding of soil, but responds well to fertility. Millet is successful on a wide variety of soils - from light sandy loam to heavy loam. High yields are obtained on chernozem and chestnut soils. Swampy, acidic, cold, heavy clay, saline soils are not suitable for millet. Millet does not tolerate high acidity and does better at pH 6.5-7.5.
Millet is characterised by increased sensitivity to soil clogging, which is due to its biological peculiarities: slow germination and slow growth in the first period of life, late tillering and light-loving. Therefore, the main requirement for the soil during cultivation is cleanliness from weeds. Specialised weed killers include various species such as chicken millet, thousand-headed millet and mice.
Depending on variety and conditions, the growing season ranges from 60 to 115 days.

1.4. Cultivation technology

Place in crop rotation. Millet gives good yields when sown as a second crop after fallow, on a layer of perennial grasses, legumes, winter crops. Potatoes and melons are considered good predecessors, sunflower, sugar beet and winter crops are somewhat worse.
Millet should not be sown on maize or before maize, as both crops are affected by maize moth. Repeated sowing of millet in the same place is undesirable due to mass reproduction of parasitic fungi (Fusarium, Helminthosporium).
Soil treatment. Particular attention should be paid to maximising weed control, conserving moisture in the soil and creating a levelled surface and seedbed.
Before ploughing carry out stubble husking or grass disking (LDG-15A, LDG-10A). On heavy soils, after perennial grasses - BDT-3,0; BDT-7,0; BDT-10. Then do mouldboard ploughing to a depth of 22-25 cm.
In zones subject to wind erosion and in areas with insufficient moisture, OPT-3-5 is used for cultivation of perennial grass beds. Deep loosening ploughs KPG-250A, KPSh-9, KPE-3,8 and needle harrows BIG-3A are used after stubble predecessors.
Spring cultivation consists of early spring harrowing and two cultivations. The first cultivation should be carried out to a depth of 8-10 cm when weeds appear, the second cultivation should be carried out to a depth of 4-5 cm before sowing millet together with rolling.
Fertilisation. An average of 30 kg of nitrogen, 14 kg of phosphorus, 32 kg of potassium and 10 kg of calcium are taken out with 1 tonne of

millet grain and 2 tonnes of straw.

Organic fertilisers. Millet makes good use of the after-effects of manure, giving yield increases from 10.8 to 32.3%.

Phosphorus and potassium are applied in autumn, nitrogen - in full calculated dose - under pre-sowing cultivation. In rows with seeds during sowing it is necessary to apply granulated phosphorus fertiliser at a dose of 15-20 kg/ha. Application of part of nitrogen fertilisers in the form of top dressing is most appropriate on wide-row crops at a dose of 15-20 kg / ha at the first inter-row cultivation. The use of nitrogen for late foliar fertilisers in the grain filling phase makes sense only to increase the protein content of grain (5-10 kg d.w./ha in the grain filling phase).

During the initial period of development, millet is particularly sensitive to phosphorus deficiencies. Before tillering, millet consumes the most nitrogen (7-8% of the total requirement) and in descending order - potassium, calcium, phosphorus.

Plants use nutrients most intensively during the period "tillering - flowering" (70% of nitrogen, 60% of phosphorus and all potassium). The greatest amount of phosphorus is assimilated during the last period of vegetation, when grain is formed and protein is accumulated in it.

Preparation of seeds for sowing. Healthy, well sorted, equalised and cleaned from impurities seeds of zoned varieties should be used for sowing. For 20-30 days before sowing seeds are subjected to air-heat heating. Seeds are treated with systemic seed dressing agents.

Sowing. Millet, sown in unheated soil, is strongly clogged by weeds that had time to rise earlier, so it should be sown when the soil at a depth of 10 cm warmed to 12-14^{0}C.

According to long-term data, the best sowing dates for Altai Krai: in Kulunda and Rubtsovsko-Aleisky zones - I decade of June; in Priobskaya forest-steppe, foothills of Salair and Altai - III decade of May - 1st five days of June.

Seeding rate for the zones of the region is differentiated as follows: for continuous crops in the Kulunda steppe - 2.5-3 million germinating grains per 1 ha; in the Aleyskaya steppe - 3.5 million germinating grains per 1 ha; in the central and eastern regions of the region - 4 million/ha; for broad row crops, respectively, 1.8-2.5; 2.0-2.5; 3.0-3.5 million germinating seeds per 1 ha.

The sowing method is row and narrow row sowing. In case of separate harvesting of such crops, losses are the lowest. On weedy soils and when there is a lack of moisture, wide-row sowing can be advantageous: double row sowing (60x15x15 cm; 45x15x15 cm) and single row sowing with row spacing of 45 and 30 cm. Wide-row sowing enables

mechanised weed control in the row spacing.
Seed sowing depth varies from 3 to 8 cm depending on the sowing date, soil moisture and nature of the soil. When sowing on light soils and in dry springs, it is increased to 8-10 cm.
Crop care. The first technique is rolling. Sometimes light perching of pearl millet crops after rains or irrigation is used.
Weed control. In the tillering phase the crops are treated with herbicides. In wide-row crops after the emergence of sprouts loosen the inter-rows and fertilise: 1st treatment - at a depth of 4-5 cm after marking the rows; 2nd - in the tillering phase at 6-8 cm; 3rd - after 10-15 days.
Harvesting. Millet ripens very unevenly, and by the time of harvesting the stalks are still juicy and green (60-70% moisture content). Millet ripening starts at the top of the panicle. This period lasts 10 days or more per panicle. The ripened grain falls off easily. Separate harvesting is best suited to millet biology.
In production conditions, mowing millet into swaths should be started when 75-80% of grains are ripe, and finished no later than when 90-95% of grains are ripe. Millet is mown with mounted reapers. Rubberised belt pads are attached to the reel slats to soften the impact on the brooms. The cutting height should be at least 12-15 cm.
Millet is threshed with a combine harvester 3-4 days after mowing, when the swaths are sufficiently dry. To prevent losses and grain collapse, the combine is retrofitted with and carefully regulated, and the threshing drum speed is reduced to 500-600 per minute.
Sometimes, especially in seed crops, a second threshing is used. In this method, the mown crop is first threshed on soft mode, as a result of which only the ripest and largest seeds are released. The unripe and small seeds remain in the brooms on the stalks, which, when coming off the straw walker, are placed in swaths for drying. On the second pass of the combine, the swaths are picked up and the dried seeds are completely threshed.

CHAPTER 7

7. *SORGO*

7.1. Importance, distribution and yield

Sorghum is a forage, technical and food crop. Grain sorghum grain contains 12-15% crude protein, fat - 3.5-4.5, nitrogen-free extractive substances - 71-82, fibre - 2.4-4.8, ash - 1.2-3.2%. Its digestibility reaches 53-85%. 100 kg of grain contains 118-130 k.u. of nutrients.

Sorghum is one of the most important bread plants in many African countries, India, China, Central Asia. The grain is used to produce flour, cereals, starch and fodder industries and alcohol production.

Sugar sorghum stalks containing 15-17% sucrose are used to produce syrup. In arid areas, it is an alternative sugar crop to sugar beet, which has a number of advantages: more resistant to diseases and pests, less pesticide consumption, environmentally cleaner production, simpler technology and lower costs. Broom sorghum is used to make brooms and is also used as a shrub crop and to create strips to protect crops from dry winds.

Sorghum leaves and stems are used to make silage. Sugar sorghum plants with succulent stems and leaves that remain green for a long time are especially good for this purpose. Sudan grass as a type of sorghum, sorghum-sudan hybrids are grown for green fodder, haylage and hay. In 100 kg of green mass - 23.5 k.e., hay - 49.2, silage - 22.0 k.e.. After mowing, sorghum grows back and the sorghum is used for green fodder or as pasture.

The centre of origin of sorghum is Africa. The plant was cultivated 3000 years BC in India and China. Currently, the global area under sorghum cultivation is about 44 million hectares, with an average yield of 1.39 tonnes/ha. Sorghum crops are widely spread in India, China, Africa, in arid areas of the USA, Europe, Australia. In our country sorghum is cultivated mainly in arid steppe areas well supplied with heat (in the Lower Volga region, in the North Caucasus). In Altai Krai, sorghum can be cultivated for silage, but this crop is not widespread, as its cultivation is connected with the supply of seeds from outside the Krai. Organisation of own seed production in the region is impossible due to the lack of heat for this crop, which has a long growing season and high heat requirements. Sudan grass is more widespread in Siberia and Altai as one of the sorghum species whose seeds mature stably in the conditions of Altai Krai. Sudan grass is grown for to produce green mass, but also as fodder for livestock.

The average grain yield of sorghum in the world is 1.5 tonnes/ha, and up to 2.5-5 tonnes/ha at variety plots. Silage yields are 20-40 t/ha, and up to 100 t/ha on irrigation. In dry conditions, the yield of sorghum hybrids is higher than that of maize.

7.2. Botanical characters and classification

Sorghum belongs to the genus Sorghum, which unites many annual and perennial species, of which the most common in culture are common sorghum (S. vulgare Pers.), gaoliang (S. chinense Jakushev), jugara (S. cernuum Host) and Sudan grass (S. sudan grass). Chinense Jakushev, djugara - S. cernuum Host and Sudan grass - S. sudanense Pers. Sudanense Pers. Among wild species of sorghum, the vicious weed humai is known. Sorghum is divided into three subspecies according to the character of panicles: the expanded (panicle) ssp. effusum Korn., the compressed ssp. contractum Korn. and the clumped ssp. compactum (Fig. 10).

Figure 10: Sorghum panicles:

1 - clumping with a straight stem; 2 - clumping with a curved stem end (jugar); 3 - corolla (overhanging) with a shortened main axis and long lateral branches; 4 - overhanging.

major axis

The root system of sorghum is arching, up to 2.5 m long and extends 60-90 cm laterally. The roots have high sucking power. Nodal and support roots are formed from underground and above-ground stem nodes. Sorghum is able to form nodal roots in a desiccated soil layer. The stem is 0.5 to 2.5 m high and up to 7 m in the tropics, parenchyma, branching, productive bushiness is 1-8. Leaves are broad - 10-25 per plant, covered with a waxy plaque, during drought the leaves curl into a tube, which reduces moisture consumption for transpiration. The inflorescence is a panicle of 1560 cm, with two spikelets at the ends of the lateral branches. One is asexual, the other male, which falls after flowering. Sorghum is predominantly cross-pollinated (up to 70 per cent). Grain is filmy or naked, rounded or egg-shaped, white, brown, brown. Weight of 1000 grains is 25-45 g. Seeds with a very short dormancy period. Grains with brown and red colouring contain a lot of tannin (tannin binder), which can reduce the quality of fodder, but in the production of alcohol and maltose suppresses putrefactive processes.

Sorghum varieties can be cross-pollinated, which makes their classification difficult. In practice, sorghum varieties are distinguished by their economic use. E.S. Yakushevsky distinguishes the following groups of varieties: 1) grain sorghum - relatively low-growing, weakly bushy, stem core is dry or semi-dry, the central vein of the leaf of an adult plant is white, the grain is easily collapsed, bare, in food varieties without tannin. It is cultivated to produce grain, groats, monofodder, pellets; 2) sugar sorghum is cultivated for its succulent stems, for sweet syrup, silage. Plants are high-growing with increased bushiness, stem core is abundantly juicy, sweet. The central vein of the leaf is green, the grain is filmy, semi-filmy, difficult to collapse; 3) broom sorghum with a dry stem core, long panicle 40-90 cm, the main axis of which is very shortened or absent, the grain is filmy, the central vein of the leaf is white; 4) grass sorghum, or Sudan grass, sorghosudankov hybrids are characterised by high bushiness and thin stems, the grain is filmy, the central vein of the leaf is green. It is cultivated for hay, haylage and grass meal.

7.3. Biological peculiarities

Sorghum is a promising crop for arid steppe and semi-desert areas, as it is the most drought-resistant plant among field crops, well tolerates both soil and air drought, continuing to assimilate when maize plants lose turgor and wither. Plants use moisture very sparingly, with a transpiration coefficient of 150-200. Seeds need a little moisture - 35% of the seed weight - to swell. In the initial period of 30 days, sorghum, as a short-day crop, develops slowly and is very sensitive to weeds. For the period of drought plants can go into anabiosis, "freeze", and when precipitation continues to grow, making good use of precipitation in the second half of summer. Sorghum is a heat-loving crop. The minimum temperature of seed germination is +12...+130C, seedlings die at frosts - 2^0C. Sorghum plants develop well at temperatures +30 ... .+350C, the sum of active temperatures during vegetation is $2250\text{-}2500^0$C. It is a light-loving plant.

Sorghum is a low soil demanding crop, it can grow both on heavy and light soils, well tolerates salinity, but prefers loose, well-warmed soils. It responds well to the application of organic and mineral (nitrogen and phosphorus) fertilisers.

7.4. Cultivation technology

Sorghum in crop rotation is placed after winter crops, leguminous crops, silage maize. It tolerates repeated sowing well, and if necessary, if it is not affected by bacteriosis, it can be grown on a permanent plot for several years.

Autumn tillage for sorghum consists of 7-8 cm stubble tilling

immediately after harvesting the predecessor and 25-27 cm winter tillage, as sorghum responds well to deep tillage, which reduces weeds and increases yields by up to 20%. On saline soils and fields subject to wind erosion, the main tillage consists of deep no-tillage loosening.
Before sowing, cultivation is carried out to the sowing depth.
Sorghum responds best to the full mineral fertiliser $N_{60}P_{60}K_{60}$ applied under the grain. When sowing in the rows, fertiliser should be applied at a dose of $N_{10}P_{10}$.
Seeds are sorted, warmed and dressed before sowing. Sowing is started when the soil warms up to 15^0C at the sowing depth. In Western Siberia it is the third decade of May. Sowing method - wide-row dotted with row spacing of 60-70 cm, distance between plants in the row 15-20 cm. Standing density when growing for grain in southern regions: 60-160 thousand pcs/ha - in the temperate zone, 40-50 thousand pcs/ha - in the arid zone, seed consumption - 10-15 kg/ha. Studies of Siberian scientists have established that sorghum for silage in Western Siberia is more appropriate to sow with row spacing of 45 cm and plant density of 350-450 thousand plants/ha. Precision seeders are used for sowing. When growing for green fodder, hay and haylage sorghum is sown in a row method with a standing density of 800 thousand pcs / ha conventional grain seeders with seeding rate of 25-30 kg / ha. Solid sowing of sorghum should be placed on weed-free fields, as weed control is difficult. The sowing depth is 4-6 cm, after which it is rolled.
If the crops are dense enough and well rooted, in the fight against weeds, as well as to destroy the soil crust conduct pre-emergence (4-5 days after sowing) and post-emergence (in the phase of 3 leaves) harrowing across the rows. It is better to use rotary hoe MV-2,1. In wide-row crops 2-3 inter-row cultivations are made. For weed control apply herbicides in the phase of 3 sorghum leaves.
In order to increase protein content and fodder value of green fodder, sorghum can be sown together with leguminous crops: soya, beans, and in dry conditions - chickpea, chinoa. Sowing 50-80 kg/ha of legumes in row crops or cross crops is used.
To increase sugar content in stems when growing sorghum as a sugar-bearing crop, treatment with sugar accumulation stimulator (paraaminobenzoic acid 0.4 kg/ha together with retardant - camposan 1 kg/ha) in the phase of the beginning of tube emergence, as well as cutting of panicles in the flowering phase is effective.
Grain sorghum is resistant to shattering, so single-phase harvesting of grain sorghum is started in the phase of full ripeness by combine harvesters with reducing the number of revolutions to 500-600 per minute. Separate harvesting of sorghum with grain moisture of more than 20% is possible with a sorghum harvester SM-2.6 or a combine

harvester.
Sorghum is harvested for green fodder and hay at the sprouting phase, for silage - at the phase of waxy ripeness of grain, sugar sorghum for syrup - at the end of waxy ripeness on a low cut, broom - at the beginning of full ripeness, while the branches of panicles should still be green. Broom sorghum stalks are harvested manually, while the remaining stalks are harvested for silage mechanised.
In young sorghum plants and awns, hydrocyanic acid can accumulate, but its content decreases with age, and hydrocyanic acid decomposes in the mass that has been wilted for 3-4 hours. In recent years, sorghum varieties and hybrids have been developed with hydrocyanic acid content within the permissible range of 20-100 mg/kg dry matter.

8. GREAT.

8.1. Importance, utilisation, distribution, yields

Buckwheat is a valuable food cereal crop. The grain contains up to 82% starch, protein - 10-12, fat - 2-4, sugars - up to 0.3, fibre - 2, ash - 2%. Buckwheat proteins are of higher quality than those of cereals, as they contain many essential amino acids (lysine - 7%, arginine - 12%). Therefore, buckwheat groats are nutritious and caloric. In addition, it contains mineral salts (Fe, P, Ca, Cu) - 21 mg/100 g, organic acids (citric, oxalic, malic), vitamins P, B1, B2, so buckwheat groats - a dietary product, especially useful in anaemia, atherosclerosis. Buckwheat fats are resistant to oxidation, as they contain a lot of vitamin E, which is an antioxidant. The iodine number is low and the oil is non-drying,

so the groats can be stored for a long time without any reduction in quality. This is of great importance when building up food stocks.

Cereal production waste, straw and chaff are used for fattening cattle mixed with other fodder.

Buckwheat is a valuable honey crop, the importance of which is increasing due to large ploughing and reduction of wild honey-bearing flora. One hectare yields up to 60-80 kg of honey. Leaves are used in the pharmaceutical industry to obtain rutin (P). The high content of cystine and cysteine determines high radiation-protective properties of buckwheat.

Agronomic importance. Late sowing date and early maturity allow buckwheat to be considered an important insurance crop. In the south, it is used as a post-cutting and stubble crop. Due to its fast growth rate and large leaf surface, it suppresses weeds well. After buckwheat, the soil remains loose due to the fine branching of roots. The agrophysical and biochemical properties of the soil are improved after buckwheat. Its root system releases acetic, oxalic, citric and other acids into the soil, which increase the solubility of phosphate, so crop residues contain more phosphate.

and potassium than stubble of cereal crops. Sometimes buckwheat stubble crops are used as green fertiliser, especially on light sandy soils.

Buckwheat crops in the world cover about 4 million hectares, of which 1.7 million hectares are in Russia. The main sown areas are concentrated in Europe. Buckwheat is inferior to cereal crops in terms of the size and stability of yields. The average yield of buckwheat in the country is 0.5-0.7 tonnes/ha. But with good agrotechnics it can give 2.5-3 tonnes/ha. Low yields are explained by higher demand for moisture, low fruit setting, simultaneous formation of leaf surface and generative organs, low agrotcchnics.

8.2. Morphological features and classification

Buckwheat belongs to the buckwheat family (Poligonaceae). The cultivated buckwheat species Fagopyrum esculentum Moench is used in production. Tatar buckwheat - Fagopyrum tataricum L. is found in the wild and is a pernicious weed (Table 20).

Table 20

Distinguishing features of common and Tatar buckwheat

Signs	Buckwheat	Tatar buckwheat
Stems	More often ribbed, reddish-green in colour	More often smooth, green.
Leaves	Heart-shaped-triangular, lance-shaped, often with a sparsely visible anthocyanine spot at the base	More rounded, more often with a well-marked spot at the base
Inflorescence	The thyroid brush	Loose brush
Flowers	Relatively large, odourless, white, pink or red, dimorphic	Small, odourless, yellowish-green, with equal length of stamens and pistil
Fruits	Relatively large, triangular, smooth	Small, triangularity weakly expressed, faces wrinkled, with longitudinal groove in the middle

Buckwheat is an annual herbaceous plant with a branching stem 50-150 cm high. The root system is tap root up to 1 m deep, but weak, the bulk of the roots are in the arable horizon. The roots age rapidly. This is the reason for the low yields and water-loving nature of buckwheat. By the end of flowering, 75% of the roots are brown. But buckwheat roots are good at absorbing phosphorus, which is unavailable to other plants.

Leaves are heart-shaped-triangular, lance-shaped, changing to arrow-shaped sessile leaves towards the top of the stem. Leaf surface is considerable, but leaf area per flower is 1.5-2 times less than in wheat. Although there can be up to 1500 flowers on one plant, only 15-20% of flowers yield. The inflorescence is an axillary shield-shaped brush with fragrant white and pink flowers. The flowers are of the five-petalled type with five corolla petals. Stamens are eight. Pistil with three columns. The flowers of buckwheat are characterised by dimorphism or heterostyly. Some plants have long stamens and short pistils, while others have the opposite. Such plants are usually about equally distributed in a population. Legitimate pollination, resulting in fruit set, occurs if pollen from long stamens falls on long pistils, or from short stamens on short pistils. Otherwise, in ilegitimate pollination, no fruit is set because the pollen does not germinate on the stigma of the pistil. The pollen is carried from plant to plant by insects. Buckwheat is a cross-pollinated entomophilous plant. Lack of insect pollinators is another reason for low yields. The fruit is a three-sided nut with a pericarp growing to the seed. The weight of 1000 seeds is 18-32 g, 15-30% filmy.

8.3. Biological characteristics of buckwheat

Buckwheat as a cultivated plant originated in the high altitude, humid regions of South Asia. This largely explains its biological peculiarities. The minimum seed germination temperature is +7...+80C. Friendly sprouts after 7 days appear at +15^{0}C. Depending on the variety and conditions buckwheat requires the sum of active temperatures from 1300 to 1700^{0}C. In spring, frosts of -1.5^{0}C damage seedlings, and at -2^{0}C seedlings die. At +12^{0}C buckwheat grows poorly, and at temperatures above +250C it is depressed, especially in the flowering phase. It grows best at temperatures close to +20^{0}C. The optimal regime for flowering is +20...250C with variable cloudiness and relative humidity of at least 60%. Under these conditions, the flowers secrete nectar well. At high temperature, nectar thickens, dries up, and bees take it reluctantly, pollination is poor, fruit setting is low.

Among cereal crops buckwheat is the most demanding for moisture. Water consumption is 2-3 times higher than that of millet and 2 times higher than that of wheat. The transpiration coefficient is 500-6°°. Seeds absorb 40-50% of the grain weight in water during germination. Water consumption from sprouting to flowering is 11%, from flowering to maturity - 89%, because along with fruiting the plants continue to grow leaves and stem. Consequently, buckwheat is more demanding to moisture in the second half of the growing season during flowering - filling, so in years with a dry second half of summer, its yield is significantly reduced. In the arid conditions of Western Siberia, where early drought is often observed, buckwheat absorbs precipitation in the second half of summer better than early spring crops.

Buckwheat is a short-day plant. Growth and development go best when the day length is 17-19 hours. If the day length is reduced to 9-12 hours, it shortens the growing season by 10-13 days. Plants grow less in height and go to fruiting faster. Variable cloudiness is most favourable.

Buckwheat is insensitive to the reaction of the soil environment and is a fairly acid tolerant crop. The optimum pH is from 5 to 7.5. This plant is quite low-demanding to soil fertility, as it increases phosphate solubility due to root secretions. It grows well on chernozems, grey forest soils, on cultivated peat bogs and sandy soils. It grows somewhat worse on sod-podzolic soils. Poorly tolerates overwatered low areas.

In Western Siberia, the most favourable for buckwheat cultivation are humid forest-steppe and foothill areas.

During the vegetation period (60-90 days) buckwheat goes through the following phenological phases: germination (2-4 days), seedlings bring seed pods to the surface (7-10 days after sowing), branching (8-10 days after seedlings), budding (10-17th day after seedlings, almost simultaneously with branching), flowering (18-28th day after seedlings). Flowering is very prolonged and can last 20-40 days. Flowers are open for 7-10 hours, and close after pollination. Then fruit formation takes place, which is also stretched for 30 days or more. The ripening phase occurs on the 25th-35th day after flowering. Maturity is marked when 75 per cent of the fruit on the plant turns white. Thus, buckwheat has a special type of growth, when all phases overlap, they cannot be distinguished from each other, so the beginning of the phase and the mass onset are usually marked.

8.4. Cultivation technology

Location in crop rotation and choice of plot. Frosts are more frequent in lowlands and on northern slopes, so it is better to choose plots on elevated southern slopes for buckwheat. Close proximity to the crops of forests, forest strips, and barks, which protect the crops from drying winds and improve the microclimate, has a positive effect on the yield. More snow and soil moisture is accumulated, air humidity is higher in summer, and there are more insect pollinators. Near the forest belt buckwheat yields are usually 2-3 c/ha higher compared to the centre of the field.

The best predecessors for buckwheat are row crops, leguminous crops, spring cereals, and in drought conditions - fallow.

Tillage. In more humid areas buckwheat responds well to mouldboard ploughing to a depth of 20-22 cm. No-tillage is used in erosion-prone areas. Snow retention and moisture closure in spring are mandatory. Before sowing, cultivation is done to the sowing depth. Buckwheat is a late sowing crop, therefore, if possible, two cultivations are carried out for a more thorough cleaning of the field from weeds.

Fertilisation. Compared to cereal grain crops, buckwheat uses more mineral substances from the soil. Phosphorus and potassium input is 2 times more, calcium input is 6 times more than that of wheat. For the formation of 1 kg of grain and the corresponding amount of straw removal is 4.4 kg of nitrogen, 2.5 kg of phosphorus and 7.5 kg of potassium. Therefore buckwheat responds well to fertilisers. It assimilates well phosphorus from phosphate meal, applied in autumn, as well as superphosphate, applied in the rows at sowing. Potassium sulphate is the best potassium fertiliser, as KC1 causes spotting. On chernozems buckwheat responds best to additional phosphorus, and on grey forest, podzol soils - on complete mineral fertilisation. Phosphorus and potassium increase nectar secretion in flowers, so bees are more willing to visit the plants and increase lakeiness.

Seeding. The period of seed formation is very long, so the seeds are not uniform in size and weight. Before sowing it is necessary to select heavier fractions on the sorting table PSS-2.5, heating in the sun or in active ventilation units, seed dressing.

Buckwheat can be sown when the soil temperature at a depth of 10 cm will be +12...+14^{0}C and there is no threat of spring frosts for the period of sprouting. In Altai, buckwheat is sown after 25 May. Sowing after 5 June is damaged by frosts in autumn. Buckwheat harvest depends on the weather by 50 per cent and is greatly reduced if flowering is caught in heat and drought. If buckwheat is sown in three dates 7 days apart, with the expectation that any one date will be in more optimal conditions, the yield will be more stable overall.

Buckwheat is sown in row, narrow-row, wide-row (with row spacing of 30, 45 cm), ribbon (15 + 45 cm). Depending on the conditions, each method has its own advantages. On more fertile, fertilised soils, clogged, prone to compaction and waterlogged soils when sown at the optimum and early dates of medium-ripening and late-ripening varieties higher yields give wide-row sowing. When sowing in wide rows, the expectation is that the plants will branch additionally. Additional branching does increase yields if there is enough nutrients. Varieties with a longer growing season are biologically better at branching. Additional inter-row cultivation with wide-row sowing is more effective on weedy areas, as well as on heavy soils, as it improves soil aeration. In the conditions of Altai Krai, varieties with shorter growing season are more often used. Inter-row cultivation requires additional costs, so buckwheat is more often sown row by row.

Cross-sown buckwheat with winter rye significantly increases buckwheat yield. Since autumn, sow rye in strips of 2-3 seed drill passes, leaving the same strips unsown. In spring, buckwheat is sown in them. The strips are placed across the prevailing winds. Winter rye grows early in spring and a greenhouse effect is created in the inter-band space, the soil is warmer by 2-3^{0}C, buckwheat sprouts more quickly and is better protected from frosts. During the flowering period of buckwheat, high-growing rye evaporates a lot of moisture, increasing air humidity, pollination and flower setting are improved. When the rye is harvested, the buckwheat is ripe, in ventilated corridors, and matures more quickly.

Seeding rate: in row sowing in steppe - 3 million/ha, in forest-steppe and foothills - 3.5-4 million/ha of germinated seeds. In wide-row sowing - 2-2.5 million/ha.

Sowing care. After sowing, roll in case of lack of moisture. Pre-emergence harrowing with light harrows or rotary harrows is done to break the crust and kill weeds in the white thread phase. On wide-row crops do inter-row cultivation at a depth of 4-6 cm with the emergence of buckwheat sprouts, then - with the emergence of weeds before flowering. Additional bee pollination increases buckwheat yield. During flowering, 2-3 bee-families are installed on 1 ha. If this is not possible, you can carry out additional pollination by dragging gauze over the crops. In this way pollen is transferred from plant to plant.

Harvesting. Buckwheat fruits form and ripen within 25-30 days. Maturation is very prolonged, accompanied by shattering of the most complete lower fruits. Therefore, a two-phase harvest is more often recommended, which is started when 2/3 of the fruits on the plants have turned brown. The rollers are picked up at a seed moisture content of 16-17%. The drum rotation speed during threshing is reduced to 500-600 rpm.

CHAPTER 9

9. *GRAIN LEGUMES*

9.1. Importance of leguminous crops, sown area and yield

Grain legumes include peas, beans, pulses, soya beans, chickpeas, chickpeas, lentils, china, lupins and peanuts. They belong to the legume family (Fabaceae). All these crops are characterised by high protein content in seeds due to symbiosis with nodule bacteria, which assimilate nitrogen from the air (Table 21). Protein contains a large number of essential amino acids (lysine, valine, tryptophan, methionine, etc.). In addition, seeds contain fats (especially abundant in soybean, peanut), minerals, vitamins A, B1, B2, C, D, E, PP, so all these plants are valuable food and fodder crops.

Table 21

Chemical composition of grain legume seeds, % of ASV

Culture	Protein	Starch	Fat	Raw fibre	Ash
Peas	20-35	20-48	0,7-1,5	5-7	2-3
Beans	17-32	50-60	0,7-3,6	2-7	3-4
Lentils	21-36	47-60	0,6-2,1	2-4	2-4
Soybeans	27-50	20-32	13-27	3-7	4-5
Beans	25-33	50-55	0,8-1,5	3-6	2-4
Chickpea	18-29	47-60	4-7	2-12	2-4
Chyna	23-34	24-25	0,5-0,7	4-5	2-3
Narrow-leaved lupine	26-36	17-39	3-5	10-18	2,9-4

One of the food problems nowadays is the lack of protein in food products. According to medical standards, a person should consume 90 g of protein per day. On average in the world this figure is 60 g, in developed countries - 90 g, in developing countries - 25 g per day. There is a deficit of protein, especially of animal origin, it is consumed 4 times less than the norm.

Animal feed also lacks protein. According to zootechnical norms, 110-115 g of digestible protein should be used per one fodder unit of protein-balanced fodder, but in fact the average amount of protein used in the country is 96 g, i.e. 87% of the norm, which leads to overconsumption of fodder and its shortage. It is necessary to solve this problem with the help of wide involvement of leguminous crops, as they contain 2-3 times more protein in the crop compared to cereals. In addition, the protein of these crops is more complete, as it contains 1.5-3.0 times more essential amino acids.

Such crops as sown peas, common beans, soya beans, large-seeded lentils have high food value. Seeds of these crops are used for food, groats, flour, added to confectionery products. Vegetable varieties of peas and beans are eaten fresh and tinned. Soy has the most valuable amino acid composition, close to animal protein, so it is added to sausages. One of the soya proteins - glycine - is able to coagulate when souring, so it is widely used to produce fermented milk products. Soybean seeds are used to produce soya oil, while the cake and meal contain up to 40% protein and are used for livestock feed.

Grains of pea, soybean, fodder beans, chickpea, chickpea, chickpea, non-alkaloidal lupine varieties are used for preparation of high-protein concentrated animal feed. 1 kg of grain of these crops contains up to 1.1-1.3 k.u. of protein.

and up to 170-250 g of protein. In addition to seeds, hay, green mass and straw of these crops are used for cattle feed. Dry green mass contains 3-8% protein, i.e. 2 times more than that of cereals.

The agrotechnical value of these crops as good predecessors is also high, as they deplete the soil of nitrogen less than non-legumes, leaving with crop residues 40-100 kg of nitrogen per 1 ha, which is equal to 10-20 tonnes/ha of manure. If legumes are used as sideral crops, the soil is enriched with biological nitrogen, which is assimilated by them in the process of symbiotic nitrogen fixation.

In world agriculture, grain legumes occupy about 13% of grain crops (Table 22). In our country, peas are the most widespread. In dry-steppe regions, chickpea and china are more important. In more southern areas, as well as in the Far East, soya is widespread, in more humid areas of the forest and forest-steppe zone - peas, fodder beans, on sandy soils - yellow lupine.

Table 22

Cultivation areas and yields of grain legumes

Cultures	Sown area, mln ha		Yield, tonnes/ha	
	worldwide	in Russia	worldwide	in Russia
Peas	6,67	0,998	1,82	1,17
Fodder beans	2,21	-	1,49	-
Soybeans	66,47	0,431	2,15	0,66
Lupin	1,42	0,015	1,14	1,03
Beans	26,36	0,004	0,66	0,85
Chickpea	11,26	-	0,74	-
Lentils	3,38	0,004	0,84	0,7

9.2. Peculiarities of the structure of leguminous crops

Seeds of legumes consist of a seed coat and an embryo. The embryo consists of two seed pods with nutrient reserves, a germinal root and a kidney. The fruit of leguminous crops, the bean, consists of two flaps, between which several seeds are attached to the stalks. When ripe, the fruit cracks at the seam and the seeds fall out. Chickpea fruits do not crack. Distinguishing features of seeds and beans are shown in Tables 23, 24.

Leaves of leguminous crops are complex, consisting of a petiole, several leaflets and a shamrock. According to leaf structure, they are divided into groups: plants with pinnate leaves (pea, bean, china, lentil), with triple leaves (soybean, bean), with palmate leaves (lupine) (Table 25, Fig. 11, 12). Plants with pinnate leaves form a seedling mainly due to the epicotyl (epicotyl) and do not bring the seed pods to the surface during germination. Deeper seed placement and harrowing before and after the emergence of seedlings are allowed in their cultivation.

Distinguishing features of grain legume seeds

View	Size, MM	Macca 1000 seeds, g	Seed shape	Seed colouring
1	2	3	4	5
Seed pea Pisum sativum L.	4-9	110-450	Round and angular, smooth and wrinkled	Pink, yellow, green
Field pea Pisum arvense L.	4-7	150-300	Round, slightly angular, with indentations	Dark, with a pattern
Fodder beans Vicia faba L.	22-30, 8-12	1-2.5kg 200-450	It's flat, roll-shaped.	Brown, black
Large-seeded lentils Ervum lens L. ssp.macrosperma	5-9	50-75	Round, flat, with sharp edges	Green, brown
Small-seeded lentils Ervum lens L. ssp.microsperma	2-5	25-50	More convex with less sharp edges	Same
Chyna sow Lathyrus sativus	7-14	100-400	Wedge-shaped, irregularly 3-4-charcoal	White, grey
Chickpea Cicer arietinum	6-10	100-400	Globular, angular with a spout	White, yellow, black
Common bean Phaseolus vulgaris	8-15	250-400	Kidney-shaped, cylindrical, spherical.	Different, with a pattern
Bean Phaseolus multiflorus Phaseolus multiflorus	17-23	700-1200	Elliptical, flattened	White, mottled

1	2	3	4	5

Lima beans Phaseolus lunatus	12-24	600-1100	Globular, kidney-shaped, moon-shaped with radial grooves	White, coloured, variegated
Sharp-leaved bean (tepary) Phaseolus acutifolius	8-10	100-140	Flattened, elliptical.	Different with radiating stripes
Golden bean (mung bean) Phaseolus aureus	3-5	30-60	Rounded-cylindrical	Yellow, green, to black.
Soybeans Glicine hispida	6-13	100-250	Globular, oval, up to elongated bud-shaped	Yellow, green, different.
White lupine Lupinus albus	10-14	250-450	Round and angular, squashed.	Cream, pink
Yellow lupine Lupinus luteus	7-10	125-150	Rounded bud-shaped	Grey speckled
Lupine narrow-leaved (blue) Lupinus angustifolius	8-12	150-180	Same	Grey with marble pattern
Lupinus polyphyllus (perennial) Lupinus polyphyllus	3-5	70-100	Oval, weakly lobed	Grey with a mottled pattern

Distinguishing features of grain legume beans

Culture	magnitude	Shape	Colouring of mature fruits	Swelling
Sown peas	Large, 3-10 grains	Straight or sickle-shaped, wide	Straw yellow	Naked
Fodder beans	Large, 3-7 grains	Long, bloated	Black	Weakly barbaric
Lentils	Length up to 2 cm, 1-2 grains	Rhombic, slightly convex	Straw yellow	Naked
Chyna sow	Small, 2-4 seeds	Wide, elongated, with two wings on the dorsal seam	Same	Same
Chickpea	Short, 2 seeds	Oval, bloated, with a spout.	Same	Densely pubescent
Common bean	Long, multi-seeded	Cylindrical, straight or curved	Straw yellow	Naked
Soybeans	Elongated, 2-5 seeds	Slightly curved	Light brown	Densely pubescent
White lupine	Large, multi-seeded	Broad, flattened, distinct.	Yellow	Same

Distinguishing features of leaves of grain legumes

View	Sheet type	Shamrocks	Leaf shape	Swelling
1	2	3	4	5
Field peas	Parene peduncles with a long tendril	Very large, with a red spot at the base	Large, rounded, oval	Naked
Sown peas	Same	There is no stain at the base	Same	Same
Lentils	Pinnate with a long tendril, multiparous	The bracts are smaller than the leaflets	Small, oval.	Same
Chyna sow	Pinnate, with tendril, unipinnate	Same	Larger, lanceolate	Same
Fodder beans	Pinnate, with a short point instead of a tendril	Small, toothed at the edges	Large, slightly oval	Naked
Chickpea	Leafless	-	Small, ovate, toothed along the edges	Densely pubescent with glandular hairs
Common bean	Triangular, petiole shorter than or equal to the leaf plate	-	Large, triangular with an elongated tip	Bare or faintly pubescent
Marigold bean	Same, but with shorter petiole	-	Same, but with a less pointed ending	Same
Sharpie bean	Similar to the common bean	-	Smaller than the common, 1.5 times smaller than the common	Same
Golden beans	Same	-	With even smaller leaves	Same
Soybeans	Triplets	-	Ovoid, oval	Strongly pubescent
Narrow-leaved lupine	Fingered, small	-	Long-linear, 7-9 in number.	Rare, appressed, on the underside only

Lupine yellow	Fingered, medium to large, 8-11 leaflets	-	Broad, large, elongate-obovate.	On both sides of the sheet
White lupine	Same, 7-9 leaves	-	Egg-shaped	Underneath, going over the edge
Perennial lupine	Same, 9-16 leaflets	-	Broadly lanceolate, acuminate at the end	On the underside of the sheet

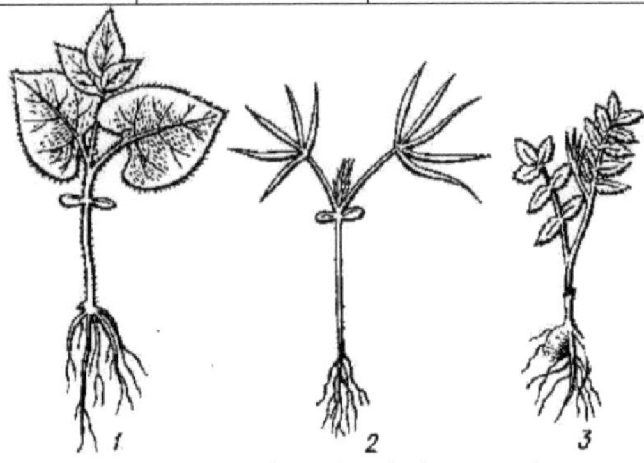

Figure 11. Sprouting of grain legume plants:
1 - with triple leaves (common bean);
2 - with palmate leaves (lupine); 3 - with pinnate leaves (chickpea)

Figure 12: Leaves of grain legumes:
1 - peas; 2 - fodder beans; 3 - chinas; 4 - lentils;
5 - chickpea; 6 - vetch

Plants with triangular and palmate leaves germinate at the expense of the hypocotyl (hypocotyl), bringing the seedlings to the soil surface. They should be sown relatively shallower, so they should not be harrowed before sprouting, otherwise there is a risk that the seedlings will be picked.

The root system of leguminous crops is rod-shaped, i.e. in the beginning a rod-shaped main root is formed, which can penetrate to a depth of 1.5

m and more, later it gives rise to lateral roots of different orders, located mainly in the arable horizon. The roots produce pinkish coloured nodules in which nodule bacteria of the genus Rhizobium settle, which assimilate atmospheric nitrogen. This improves the nitrogen nutrition of legumes and enriches the soil with nitrogen.

According to the stem structure of grain legumes, the following groups can be distinguished:

1) having an erect, relatively lodging-resistant stem (soya, beans, lupin, chickpea);
2) having a lodging stem (peas, chickpeas, lentils).

Plants of the second group have pinnate leaves in which the upper leaflets are reduced into tendrils.

The flowers of plants of the legume family are irregular, moth-like. The corolla consists of five petals. The upper, unpaired, larger than the others is a sail, the two lower ones are fused at the bases, forming a boat, the two lateral free ones are wings. The stamens are ten, nine of them fused, surrounding an elongated one-nested ovary with several seedpods.

As plants develop and new vegetative organs are formed, flowers or inflorescences are formed. In most species, single flowers are formed or 2-3 flowers are formed in leaf axils, only lupine forms an apical brush. This type of growth, when it continues in height at the same time as budding and flowering, is called non-determinant and leads to unfriendly, prolonged ripening.

Nowadays, pea forms with a shortened, thickened stem, in which the flowers are located mainly in the upper part of the stem (boll forms) are widely used, which causes their more friendly, simultaneous ripening. Many modern varieties are based on mutant forms of peas in which the leaves have mutated into whiskers. Such plants are held together by numerous whiskers and do not lodge.

Leguminous crops are mainly self-pollinators, but in fodder beans, white, yellow and perennial lupine, along with self-pollination, significant (up to 50%) cross-pollination is observed. In soya, beans, chickpea, chickpea, along self-pollination, insignificant cross-pollination is observed.

Grain legumes have the following phenological growth phases: germination, sprouting, stem branching, budding, flowering, bean formation, ripening, full maturity.

9.3. Biological characteristics of grain legumes

Temperature, moisture, light regime have a decisive influence on the formation of grain legume crop yield.

Most of these crops are long-day plants, species of temperate climatic zones, such as peas, beans, lupines, chinens, lentils, and chickpeas. They are less demanding for heat (Table 26).

Soybeans and beans are short-day plants, but early maturing varieties of these crops are photoperiod-neutral. These plants are of more southerly origin and are more heat-loving.

Table 26

Heat requirements of grain legumes

Culture	Seed germination temperature,0C min/opt	Optimum temperature for the growth period и of development, 0C	Sum of active temperatures, 0C	Frost damage temperature in the phase of sprouting/growing, ^{0}C
Peas	1-2/6-12	16-20	1200-1600	-6...-7/-3...-4
Beans	3-4/9-12	16-20	1400-1900	-5...-6/-3...-4
Lentils	2-3/6-12	16-22	1200-1600	-6...-7/-3...-4
Chyna	2-3/6-12	16-22	1800-1700	-6...-7/-3...-4
White lupine	2-4/9-12	16-20	1700-2300	-4...-5/-3
Narrow-leaved lupine	2-4/9-12	16-20	1300-1600	-5...-6/-3...-4
Chickpea	3-4/9-12	18-21	1300-1700	-8...-9/-3...-4
Soybeans	7-9/15-18	18-25	1700-3200	-2...-3/-2
Beans	6-10/15-18	18-25	1500-1900	-1/-2

For leguminous crops, higher levels of
temperatures for the period of seed filling and ripening.

In general, grain legumes are more water-loving than cereals. A lot of water (100 to 120% of seed weight) is required for the period of seed swelling and germination, as the seeds contain a lot of protein.

These crops have high transpiration coefficients (500-600), which is higher than that of cereal crops. Young plants grow slower, especially in short-day crops, and therefore use winter-spring moisture reserves less efficiently. The maximum moisture demand of grain legumes is observed during flowering and fruiting. Symbiotic nitrogen fixation is ineffective in case of moisture deficiency,nodules are not formed,
yields are significantly reduced. Optimum soil moisture is not less than 70% of NV. The most moisture-loving crops are peas, beans, soya beans, lupine, while the most drought-resistant are chickpeas and chinens. Lentils and beans are relatively drought tolerant.

Leguminous crops are characterised by rather high removal of nutrition elements (Table 27).

Table 27

Nutrient inputs per 1 tonne of seed
and related by-products, kg

Cultures	Takeaway

	N	P2O5	K2O	Total
Sown peas	50	16	24	90
Field peas	45	20	17	82
Chickpea	52	21	49	122
Fodder beans	52	20	44	116
Beans	53	22	29	104
Chyna	58	16	30	114
Lentils	59	20	28	107
Narrow-leaved lupine	67	19	43	129
Lupine yellow	68	19	42	129
Soybeans	72	23	38	123
On average	58	19	23	110

Nitrogen excretion averages 58 kg per 1 tonne of seed, compared to 34 kg for cereals. It is believed that legumes assimilate 2/3 of nitrogen from air and the rest from soil. Therefore, under good conditions for nitrogen fixation these plants are less demanding for nitrogen nutrition, but in case of inefficient nitrogen fixation these plants switch to nitrogen nutrition only from soil, then the yield is significantly reduced. Phosphorus and potassium removal is also high, so these crops are more demanding for these nutrients also because they increase the efficiency of symbiosis.

The trace elements molybdenum and boron are of great importance for legumes. Molybdenum is part of the enzyme nitrogenase, which binds air nitrogen, and is important for biological nitrogen fixation. Boron promotes better development of the vascular conducting system, which transports assimilants from leaves to nodules on roots and from leaves to fruit. Boron significantly increases fruit setting.

In general, neutral soils of medium mechanical composition containing enough phosphorus, potassium, calcium are preferable for grain legumes. Legumes have different requirements to the reaction of soil solution. The most acid-tolerant are yellow lupine and perennial lupine, they form an effective symbiotic apparatus on soils with pH 4.5-5.0, can grow on sandy soils. Field peas and narrow-leaved lupine grow well on sandy slightly acidic (pH 5.5) soils. On slightly acidic (pH 5.5-6.0), but more fertile soils can grow sowing peas, fodder beans. Soybean, white lupine prefer fertile soils with pH 6.0-7.0. Beans, chickpeas, chickpea - soils with pH 6.5-7.5. Beans, chickpeas and chickpeas tolerate slightly saline soils well.

9.4. Main features of cultivation technology leguminous crops

Place in crop rotation. In order to comply with phytosanitary requirements, grain legumes should not be returned to the same place earlier than after 3-4 years. In crop rotation they are placed after cereals and after row crops. They should not be placed after perennial leguminous grasses and grain legumes, as they have common pests and pathogens.

Grain legumes themselves are good predecessors for cereals and row crops, as they deplete the soil of nitrogen less than other crops.

Peculiarities of fertiliser application. Plants of the legume family respond well to phosphate and potassium fertilisers. The more acid-tolerant the crop is, the lower its limit of phosphorus availability. Acid-tolerant yellow and blue lupine grow well at low phosphorus content , the lower limit is 50 mg/kg of soil.

Soybeans, peas, beans respond well to liming on acid soils and have a lower phosphorus limit of 150 mg/kg, beans - 200 mg/kg.

Under optimal conditions of nitrogen fixation (good supply of moisture, phosphorus, potassium, trace elements B and Mo, soil aeration, presence of specific bacterial strains, neutral reaction of soil medium), nitrogen nutrition is provided by symbiosis for 2/3 and from soil for 1/3. In this case, the natural fertility of the soil can provide a yield of up to 2.5 tonnes/ha of seeds. Application of $P_{60}K_{60}$ in such conditions increases the yield up to 3 tonnes/ha. Application of additional nitrogen to the soil may be economically unfavourable, as plants switch to feeding on more available mineral nitrogen. Biological nitrogen is not assimilated, nodules are not formed for a long time, and yield increases are not observed. To obtain a yield of more than 3 t/ha, it is necessary to additionally apply nitrogen, but to overcome the antagonism of auto-trophic and symbiotrophic nutrition, it is better to apply nitrogen in the form of foliar fertiliser in the phase of budding - flowering. Mineral nitrogen assimilated through the leaf surface does not conflict with the assimilation of biological nitrogen from the atmosphere.

Under conditions unfavourable for nitrogen fixation, nodules do not form or are ineffective, then nitrogen fertilizers are effective and used depending on soil fertility and planned yield.

It is better to apply lime under the predecessor, so that it has time to neutralise acidic soil. To reduce the pH by one unit, 10 tonnes of lime should be applied per hectare. Organic matter applied directly to legumes with unstable stems causes the risk of high lodging and crop overgrowth to the detriment of fruit formation. Plants with stable stems can be fertilised with 20 t/ha of organic fertiliser.

Tillage. The main tillage for grain legumes is the same as for cereal grain crops, including husking after harvesting stubble predecessor and deep autumn ploughing in 3 weeks. In areas prone to erosion, soil-protective flat tillage after harvesting the predecessor KPSh-9 at 8-10 cm, deep loosening KPG-2-150 at 22-25 cm. After row crops on clean fields ploughing is replaced by loosening. In winter, snow retention is mandatory. In the spring do harrowing to close moisture and soil levelling plough harrows or, if possible, using special equipment (GN-4 - for levelling ridges and furrows, VPN-5,6 or VP-8 - for general levelling). This technique promotes uniform seed placement during sowing. On the levelled field sprouts appear at the same time, there is no adjustment, plants develop evenly, mature in a friendly manner. In addition, more reliable work of stalk lifters is ensured, and yield losses from uncut beans are eliminated. These conditions are essential for more progressive direct harvesting.

For early-sown crops, one pre-sowing cultivation to sowing depth is carried out. For late-sown short-day crops - two cultivations with a break of 7-10 days to ensure higher weed clearance.

Conditioned seeds are used for sowing, dressed 3-4 weeks in advance to prevent diseases. Effective preparations are used, e.g. Fundozol (3 kg/t seeds), Tachigaren (1-2 kg/t) in PSh-5, Mobitox machines.

Seed inoculation, i.e. treatment with microbiological preparations (nitragyn or rhizotorphin) to activate symbiotic fixation of atmospheric nitrogen by nodule bacteria is very effective for legumes. Consumption of the preparation - 200 g per hectare rate of seeds. Inoculation is done on the day of sowing without access to direct sunlight, it cannot be combined with dressing, but can be done simultaneously with seed treatment with microelements. When sowing on acidic soils is effective molybdenum, 25-50 g of ammonium molybdate per 1 kg of seed. When sowing on freshly limestone neutral soils seeds are treated with boric acid - 25-50 g / 1 kg of seeds. This seed treatment increases yield by 0.2-0.3 tonnes/ha.

If more than 20 per cent of the seed lot is hard-shelled, scarification is necessary. Hard-seededness is related to the water resistance of the seed coat and is observed in some crops of the legume family, especially when the seeds mature at high temperatures and sudden loss of moisture from the seeds. In this case, the seed scar is blocked, and moisture cannot get to the embryo, the seeds do not swell and do not germinate, although the embryo is quite viable. Seeds are scarified in special scarifiers or clover grinders. This breaks the integrity of the seed coat and allows moisture to reach the embryo and the seeds begin to germinate.

Yields depend largely on the correct choice of sowing dates. Long-day cold-resistant plants (peas, beans, chinens, lentils, chickpeas), whose seeds begin to germinate at a temperature of +2...+40C, and sprouts well withstand frosts, it is better to sow early sowing dates at the physical ripeness of the soil and its warming at the sowing depth to $+5^0C$ in the very beginning of May in the conditions of Altai Krai. Late sowing reduces yields by 15-20%, as the topsoil loses moisture, and all leguminous crops consume a lot of moisture for seed swelling (100-120% of seed weight). In case of late sowing of such crops, ripening occurs in a colder period and it is delayed, plants are more affected by diseases (powdery mildew), aphids, increased infestation with late weeds.

Short-day crops of southern origin (soya, beans) are more heat-loving. Their seeds begin to germinate at a temperature of at least +10^{0}C, sprouts poorly withstand frosts, so these crops are sown at a later date: beans - at the end of May, and soybeans - starting from 15-20 May for the conditions of Altai Krai, when the soil warms up to +10^{0}C, then there is no threat of frosts for the period of sprouting. Seeding coefficients are given in Table 28.

Table 28

Seeding rates and sowing methods for leguminous crops

Culture	Method of sowing	Seeding rate, mln/ha	Weight of 1000 seeds, g	Seed purity, % (PCT)	Germination, % (PCT)
Sown peas	Row, narrow row	1-1,4	150-250	97	87
Fodder beans	Row, wide-row	0,4-0,7	200-450	98	85
Chickpea	Row, wide-row	0,6-0,8	160-220	98	85
Coarse-seeded lentils	Row, narrow row	2-2,5	55-65	98	87
Fine-seeded lentils	Same	2,5-3	25-30	98	87
Narrow-leaved lupine	Private	1,1-1,2	150-180	95	80
Lupine yellow	Private	1,1-1,2	125-150	95	80
White lupine	Row, wide-row	0,6-0,8	240-450	96	80
Soybeans	Wide-row	0,4-0,6	100-200	95	80
Beans	Wide-row	0,3-0,5	200-400	98	87

In the steppe, if there is a lack of moisture, the seeding rate is reduced by

20-30%. In case of wide-row sowing, the seeding rate is reduced by 30% compared to the row method.

Crops that grow relatively quickly in the first period are sown in row and narrow rows. These are all long-day crops of early sowing period. Soybeans and beans, as short-day plants, grow slowly in the first period, worse tolerate mutual shading, are strongly choked by weeds, so they are traditionally sown in a wide row with row spacing of 45-60 cm. They are also used in two-line wide-row sowing with 15 cm spacing between rows.

Row sowing is carried out with grain drills. All leguminous crops have large seeds, and in order not to traumatise them during sowing, it is necessary to sow with minimum gear ratio and maximum length of the working part of the seeder spool. Wide-row sowing is carried out by precision seeders (SUPN-6, SPCH-6, SKNK-8, STV-12, SST-12A with the device STYA, vegetable seeders SON-2,8), grain seeders with anchor coulters SZA-3,6, SZ-3,6).

It is not recommended to sow crops that bear seed pods on the surface (soya, beans, lupine) deeply. The sowing depth should be 5-6 cm. Crops that do not bear seed pods on the surface (peas, chickpeas, lentils, beans, chickpeas), if necessary on lighter and drier soils can be sown deep (6-8 and up to 10 cm), and on wet and heavy soils - to a depth of 5-6 cm.

Crops with unstable stems, such as peas, especially in areas of sufficient moisture have a large vegetative mass and lodge. This makes harvesting difficult, so peas are often sown in a mixture with oats, barley, wheat, sowing 20-30% less than the norm of cereals and 50-60 kg/ha of peas. But in this case, pea yields are significantly reduced. If moisture is not enough, when growing peas according to intensive technology, it is better to sow them in pure form, which allows you to choose the right plant protection system and fertilisers.

Plant care. After sowing, the soil is rolled with ring-spur rollers (3KKSH-6A). This is especially important on light, quickly drying soils. If the soil is moist, it is not necessary to roll the soil. In the fight against annual weeds harrowing is carried out before sprouting 4 days after sowing and after sprouting in the phase of 2-4 leaves until the appearance of tendrils across or diagonally to the rows. Harrow tines are directed with the mown side forward, use caterpillar tractors with less pressure of the undercarriage on the soil, speed - 6-8 km/hour for the first and 4-5 km/hour for the second harrowing. This destroys the soil crust, improves soil aeration, destroys 60-80% of annual weeds. Post-emergence harrowing is better to be carried out in the daytime, when the turgor of plants is disturbed, then they are less traumatised. When planning harrowing before and after sprouting, the seeding rate is increased by 10-15%.

In case of shallow sowing, harrowing of crops that bring seedlings to the surface before sprouting is not recommended.

On legume crops it is possible to apply herbicides both soil (Prometrine, Lenuron - 1.5-2 kg/ha with incorporation into the soil) and during vegetation (Bazagran - 2.5 kg/ha, Fusilat-Forte - 1-1,5 l/ha against cereal weeds and Agritox - 0.5 l/ha against dicotyledonous weeds) in the phase 3-5 leaves of peas, when the maximum wax coating on the leaves, and cultivated plants are resistant to the herbicide. Herbicides are applied with boom sprayers OPSh-15, OP-2000. The use of herbicides is especially important for crops that develop slowly in the beginning of crops, in particular soybean.

On wide-row crops carry out 2-3 inter-row cultivation with cultivators KRN-4,2, timing them to the mass appearance of weeds no later than the budding phase.

In case of mass appearance of pests (e.g. aphids - 20 pieces per plant, pea weevil) it is necessary to spray the plants with insecticide.

Harvesting. Most grain legumes ripen unevenly, with the lower beans ripening first, followed by the upper beans. When ripening, they crack, especially in unstable weather after alternate moistening and drying. The lower beans with the most complete seeds have low attachment. Large vegetative mass and unstable stalks lead to lodging. All this makes harvesting more difficult and therefore separate harvesting is more often recommended. Plants are cut into swaths when 65-70% of beans are browned, when the filling is over and seed moisture is 30-35%, by bean cutters ZhRB-4,2, ZhSB-4,2. Cutting height is 5-6 cm. The leguminous stalks are mowed across the lodging or at an angle to the lodging. It is better to mow in the morning or evening, when the moisture content of beans is higher and they are less likely to crack. The swaths are picked up in 3-4 days at a seed moisture content of 16-19% by combine harvesters equipped with a conveyor copy picker. It is important to adjust the threshing machine, reducing the revolutions of drum to 400-500 rpm for dry crop and 500-700 rpm for wet crop. The gaps between the drum whips and the deck slats are 20-25 cm at the inlet and 8-13 mm at the outlet (less for wet crop, more for dry crop).

As more advanced technologies are learnt and under certain conditions, direct harvesting is acceptable:

- for crops and varieties whose beans do not crack (chickpeas, non-cracking pea varieties);
- for non-legumes (soya, chickpea, moustached or semi-moustached pea varieties, varieties with short thickened stems);
- with early seed ripening (varieties with a more determinant type of growth, short-stemmed);
- on weed-free, levelled fields;
- in steppe areas, with rapid ripening in drought;
- with preliminary desiccation of crops with reglon (2-3 kg/ha) at the beginning of yellowing at seed moisture 45%, 7-10 days later harvesting is carried out directly;
- equipping the combine with stalk lifters.

Harvesting is directly started when the fruit is 90% browning (yellowing) and seed moisture is 18-19%.

CHAPTER 10

10. CORNEPLOODS

10.1. Importance, chemical composition, yields

Root crops include sugar beet (Beta vulgaris L. v. saccharifera), fodder beet (Beta vulgaris L. v. crussa) belongs to the Chenopodiaceae family, fodder carrot (Daucus carota L. ssp. rapifera) to the Apiaceae family, rutabaga (Brassica napus L. ssp. rapifera), turnip (Brassica rapa L. ssp. rapifera) to the Apiaceae family.) belongs to the celery family (Apiaceae), rutabaga (Brassica napus L. ssp. rapifera), turnip (Brassica rapa L. ssp. rapifera) belong to the cabbage family (Brassicaceae).

The main sugar-bearing technical crop in Russia is sugar beet.

Root chemical composition: water - 75%, dry matter - 25, including sucrose - 17.5, fructose and glucose - 2.5, protein - 1, fibre - 2.5, pectin substances - 2.5, ash - 0.6%. Modern sugar beet varieties contain 15-19% of sucrose, and at high yields (40-50 tonnes/ha) sugar collection per hectare will be 7-8 tonnes/ha. Wastes of sugar beet production are molasses and cake. Treacle contains up to 60% sucrose, it is used to produce alcohol, yeast, citric acid. Beet pulp (squeezed beet shavings) contains 15% of dry matter and is a good fodder. 100kg of dry cake contains 80kg of feed units. Defecate contains up to 50% lime, 1.5% organic matter and is a good fertiliser, especially on acidic soils.

In world farming, sugar beet is sown on about 8 million ha, mainly in Europe, China, Russia and Ukraine. In Russia, the sown area is about 1 million hectares (Central Central Central Zone, Krasnodar Territory, Stavropol Territory, Western Siberia). The average yield in the world is 32.8 tonnes/ha, in the country - 15-16 tonnes/ha. In advanced farms, yields of 30 to 50 tonnes per hectare are obtained.

Fodder root crops - fodder beetroot, carrots, rutabaga, turnips - are juicy, easily digestible, lactic forage. In 100 kg of root crops contains 7-14 k.u., due to high yield (50 t/ha of root crops + 20 t/ha of haulm) yield of fodder units - 8.5-10 t/ha, which is 3 times higher than that of cereal crops. Fodder root crops contain 15-20% dry matter, including 10-12% sucrose, vitamins C, B_1, B_2, PP, carotene.

10.2. Peculiarities of the structure of root plants

All root crops are biennial plants. In the first year, a root rosette of leaves and a thickened root ball grows. In the second year, stems with flowers and fruits form from axillary buds on the head of the root crop.

Not only seeds proper, but also fruits can often be used as seed (Table 29).

Table 29

Seed for root crops

Culture	Fruit or seed	Shape	Magnitude,	Surface	Colouring

			mm		
Beetroot	The fruit is a nut or copepod	Round-angled	2-6	lumpy	Yellow-brown
Carrots	The fruit is a double seed	Elongated ovate	3	Ribbed with fine needles	Brown
Brukva	Seeds	Globular	Up to 2	Smooth	Black
Turnips	Seeds	Globular	Up to 2	Smooth	Brown to black

The fruit of the beetroot is a nut, covered with a loose, woody pericarp, on top of which, under a flat cap, there is a horizontally lying seed in a shiny shell. The embryo of the seed is coiled in a ring around the perisperm with a reserve of nutrients. The embryo consists of two seedpods, a kidney between them, a subseedlet and a germinal root.

There are biological forms of beetroot: multi-seeded and single-seeded. In multiseeded beetroot, flowers on the peduncle are gathered in whorls of 2-5 pieces, when forming fruits, they are joined by pericarp, forming a copepod of 2-5 fruits.

Single-growth beetroot has flowers arranged one at a time, producing single fruits. Single-growing beet varieties and hybrids have been developed relatively recently. Their introduction into production has made it possible to mechanise all beet cultivation operations.

The fruit of rutabaga and turnip is a pod, that of carrot is a two-seeded pod.

When the seeds are exposed to favourable conditions, they begin to germinate. During germination, these crops bring the seedpods to the surface, with further development the first true leaf appears, then the second, and so on. (Table 30, Figure 13).

New leaves are formed throughout the growing season, with the young leaves emerging in the centre of the leaf rosette and the old leaves being pushed to the periphery of the head. Beetroot can form 50-70 leaves. At the same time as the leaf surface, a root crop is formed.

Table 30 Distinguishing features of seedlings and true leaves of root crops

Culture	Seed leaves	First true and subsequent leaves				
		plate	mould	top quality	colouring	wax deposit
Beetroot	Long lanceolate	Solid	First oval, then heart-shaped.	Smooth	Green	No
Carrots	Long linear	Mno-go-krat-noras-sectioned.	-	Smooth or with hairs	Same	Same
Brukva	Oval with a notch at the	Slightly dissected.	Lyre-shaped	Smooth	Dark green	Got it

	end					
Turnips	Same	Same	Same	Lowered	Light green	No

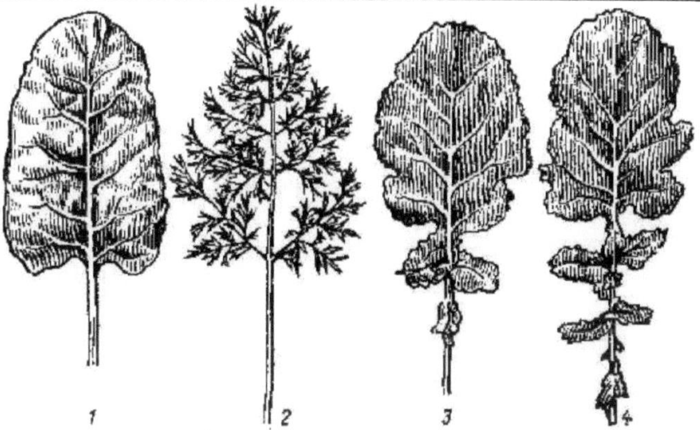

Figure 13: Leaves of rootlets:
1 - beetroot; 2 - carrots; 3 - rutabaga; 4 - turnips

A root crop consists of a head, a neck and the root itself. The head is a stem-like formation, the upper part of the root crop, which bears the leaves, develops entirely above the ground. The head is heavily woody and has the least amount of sugar. The cotyledon is formed from the subseed (hypocotyl), is cylindrical in shape, it bears neither leaves nor lateral roots. The neck, as well as the root is a full-fledged part for technical and fodder purposes. The root proper is the lower conical part of the rootlet, on which lateral roots are formed (Table 31). Rutabaga root flesh is more dense, poorly translucent even on rather thin cuts. Turnips have more friable flesh, vascular-conducting rings are well translucent on thin sections.

In turnip, rutabaga and carrot, two rings formed by vascular-conducting bundles can be seen on the transverse section of the root, which is associated with the primary and secondary structure of the root. In beetroot, 10-12 rings are formed due to the activity of several successive cambial rings forming vascular-conducting bundles. Between the rings is formed parenchyma tissue, in the cells of which the bulk of sugar is deposited.

Table 31

Distinguishing features of root crops of different species

Signs	Beetroot	Carrots	Brukva	Turnips
Location of lateral spines	On both sides of the root in two grooves	In four sparse vertical rows	Over the entire surface of the root proper	On the tap root
Root shape	Diverse	Long	Rounded	Long, rounded, conical.

Colouring of the underground part of the root	The sugar one is white, y fodder - yellow, orange, pink	White, orange, red	White, yellow	White, yellow
Colouration of the above-ground part of the root	The sugar one is white, y fodder - grey-yellow, red-violet	White, orange, green	Green, purple	Green, purple
Root pulp colouring	In sugarcane - white, in fodder - white, sometimes with pink rings, rarely yellow	White, orange, red	White, yellow	White, yellow
The flavour of the root	Sweet	Spicy	Reddish, sweeter.	Redecorated

10.3. Biological characteristics of root crops

The cycle of individual development of root crops lasts two years. In the first year, sugar beet vegetation duration is up to 150-170 days (120-130 days in Siberia), fodder beet - 25-30 days shorter than sugar beet, rutabaga and carrots - 110-120 days, turnip - 70-110 days. In the second year - 90-120 days.

In the first year, a thick root ball with a root rosette of leaves is formed, but there are variations in the form of flowering plants in the first year and "stubborn" plants in the second year. When the beetroot goes through its full development cycle in one year, flowers and fruits, it is called "flowering". Flowering root crops become woody and are unsuitable for processing. Causes of blossom blight: early sowing in a cold prolonged spring, long daylight hours. Beetroot, as a long day plant, shortens the growing season on a long day with a long cold period and may already go into flowering in the first year if the varieties are not adapted to a long day. Rutabaga and turnips, as more cold-resistant crops, are less prone to flowering than beetroot, while carrots require a long period of low temperatures (100-140 days) for flowering, so they do not flower even in winter sowing.

Stubborn or "bachelors" are plants that do not flower in their second year. This reduces the yield of the seedlings. Causes are increased temperatures during early harvesting of mother plants or during storage, drying of root crops, shallow planting in the second year.

Warmth requirements. All these crops are long-day plants, as species were formed in temperate latitudes, so they are relatively cold-resistant. The minimum seed germination temperature is +3...+40C, viable sprouts appear at a temperature of +6...+7 on the 10th-15th day, and at the optimum temperature for the sprouting period of +15...+180C - on the 67th day.

In the seedling leaflet phase, seedlings may be damaged by frosts of -3^0C, but in the phase of the first pair of true leaves, cold tolerance increases and the plants withstand frosts of -4^0C. The rootlets of adult plants, when dug up and lying on the ground uncovered, are damaged by frosts of -3^0C and become unsuitable for storage. Root crops of adult plants dug up and lying uncovered on the ground are damaged by -2^0C frosts and become unsuitable for storage. This is especially dangerous for mother plants.

The optimum temperature for assimilation in beetroot is 120...1250C, in rutabaga and turnip - +15...+200C. Rutabaga and turnips in the more southern regions do not tolerate heat and lack of moisture. In autumn, the vegetation of root crops stops at a temperature of

+2...+4^0C, and sugar accumulation stops at +60C. The sum of active temperatures required in the first year for beetroot is 2200-24000C, for rutabaga and turnip - 1500-18000C.

Moisture requirement. Sugar beet is a relatively drought-resistant crop. A strongly developed root system up to 2-3 metres deep allows water to be extracted from great depths. Having a long growing season, sugar beet effectively utilises precipitation in the second half of summer. A rather low transpiration coefficient of 400 allows for more economical use of moisture, especially at a high level of agrotechnics. At the same time, beetroot has rather high requirements for moisture supply. During the germination period it requires a lot of water - 150-170% of the seed weight for swelling and germination due to the presence of a loose pericarp. The greatest amount of moisture is consumed during the period of enhanced root growth in July - August. When there is not enough moisture for this period, a smaller root crop is formed, but with a higher sugar content as a rule.

Root crops form a large amount of dry biomass, water consumption per 1 ha is much higher than that of cereals, so all agro-techniques for moisture storage are of great importance.

Fodder root crops are less drought-resistant than sugar beet, as they have less powerful root system up to 1-1.5 m deep, so it is better to place them in low relief places.

Soil and nutrient requirements. Root crops are crops of loose and fertile soils. Beetroot is the most demanding of soils; it needs soils with a deep arable horizon; it grows best on chernozems, grey forest loamy soils rich in humus. Soils of lowlands and floodplains, meadows, dark chestnut soils are suitable for it, moist, deep

cultivated fertile sod-podzolic soils. Neutral or slightly alkaline reaction of soil solution is most favourable for beetroot. It can adapt to slightly saline soils and even forms on such soils higher sugar content. Beetroot should not be planted on heavy clay, waterlogged, poor sandy and stony soils.

Carrots grow in a variety of soils, with an optimum pH of 5.5-7.0.

Rutabaga prefers cohesive soils with good water-holding capacity and can be successfully grown on heavy soils that are overwatered, but does not do well on sandy soils. Turnips grow well on light soils, heavy soils are not suitable for them. Both for rutabaga and turnips prefer slightly acidic soils with pH 6.0-6.5, they satisfactorily withstand increased acidity of pH 4.5.

High demanding to soil fertility is caused by high removal (Table 32).

Table 32

Nutrient inputs, kg per 1 tonne of root crops and related by-products

Culture	N	P2O5	K2O
Sugar beet	4-7	1-3,5	5-7,5
Fodder beet	2,5-3	0,9-1	4,5-5
Carrots	3,5	1,5	7
Brukva	4	2,5	7,5
Turnips	2,5	1	3,8

Plants need nitrogen throughout the entire growing season, but especially at the beginning of growth. Nitrogen deficiency causes leaves to turn light green, turn yellow early and die back. Excess nitrogen lengthens vegetation, enhances leaf growth to the detriment of root growth, reduces sugar content, while the content of non-protein nitrogen increases in root crops, which reduces the sugar yield of the plant. In addition, the storability of mother and fodder root crops deteriorates, drought resistance, resistance to diseases and pests is reduced.

Phosphorus is necessary especially in the first phases of growth, has a positive effect on the root system, increases sugar content by 0.2-0.3%.

Root crops are potassium-loving plants; the potassium yield of sugar beet is 3 times higher than that of cereal crops. Potassium significantly increases sugar content by 0.3-0.6%, drought resistance, storability, resistance to diseases and frost.

Calcium deficiency causes variegation, borax causes heart rot.

Peculiarities of fertiliser application. Application of organic matter under ploughing provides high yield increase of sugar beet. 1 t/ha of manure gives an increase of 0.5 centners/ha of root crops in the steppe and 2.5 centners/ha - in the forest-steppe. Recommended rates of organic matter application: in steppe - 20-30 tonnes/ha, in forest-steppe - 30-40 tonnes/ha.

In modern economic conditions, double, triple doses of fertilisers for sugar beet do not pay back the yield gains obtained. A greater effect can be obtained by combining the application of organic matter under the predecessor and moderate rates of mineral fertilisers for sugar beet (60-90 kg d.v./ha). The application rates of phosphorus and potassium are slightly increased compared to nitrogen (by 15-20%) in order to ensure higher sugar content and root quality.

Fertilisation by spraying plants before the leaves close together in the inter-row with humic fertilisers with microelements, Nutrivant plus with adhesive, Aquamarine and others are effective in providing good plant protection against stress, growth stimulation and an additional source of nutrition.

Phases and periods of growth and development of beetroot. In the first year of life, the following growth phases are observed: sprouting (fork phase), 1-, 2-, 3-rd pair of true leaves, leaf clasping in rows, leaf clasping in inter-rows, leaf unclasping in inter-rows.

In the formation of the root crop in the first year of life, the plants go through the following periods.

Sowing - sprouting (10-15 days). If this period increases, especially in cold, wet springs and when crusts form, plants are affected by root-knot disease (this is a disease that affects plants under the action of pathogenic soil microorganisms), it is necessary to provide effective plant protection, to destroy the soil crust.

Sprouting - 3rd pair of leaves. At this time there is a transition from the primary structure of the root to the secondary one, the so-called "moulting" of the root, the primary bark is destroyed, which is manifested in the appearance of longitudinal cracks in the underground part of the plant, the root begins to grow in thickness. By the end of this period it is necessary to complete the formation of plant density, thinning, if necessary, otherwise there is a "run-off" of the root, that is, in a thickened state of plants form a thin root crop of irregular shape.

A period of increased leaf growth from the 3rd pair of leaves to leaf closure in the inter-row.

Enhanced root growth and sugar accumulation before harvesting.

10.4. Technological methods of cultivation sugar beet

Sugar beet is one of the most costly crops, the yield of which largely depends on soil, climatic and technological conditions.

At present, it is rather conventional to distinguish extensive and intensive technologies. Extensive level involves the cultivation of plastic, the most adapted to local conditions varieties-populations without fertilisers, using old-generation domestic machinery, without protection means or with occasional use of them. Intensive technologies involve the cultivation of high-yielding hybrids with high yield potential, which requires the use of fertilisers, integrated plant protection system, the use of more advanced imported machinery or its domestic analogues. With a combination of high energy and resource intensity, as well as appropriate natural resources in Altai Krai in the zone of arid and moderately arid kolok steppe can be expected 23-30 tonnes/ha, in the forest-steppe - 30-35 tonnes/ha of sugar beet root crops.

The technology of sugar beet cultivation in different soil and climatic zones has its own peculiarities. Sugar beet should be placed in crop rotation with return to the same place not earlier than in 4 years in order to create optimal water, air, food regime, as well as good phytosanitary condition of the field and clearing it of weeds.

In the steppe zone of Altai, sugar beet is traditionally placed on clean fallow. In the forest-steppe, where the sum of precipitation for the year is not less than 500 mm, it is better to place sugar beet on winter or spring wheat, going on fertilised fallow. In the forest-steppe, especially on heavier soils, according to the data of the Biysk experimental breeding station, beet sprouts sown on pure fallow, suffer more from soil crust and root-knot, as after several treatments in fallow the soil is prone to swelling and compaction. Inclusion of sideral crops in beet crop rotations as a fallow and as an intermediate crop after early harvested predecessor enriches the soil with organic matter, cleans the soil from nematodes. The use of a sideral crop is also an additional means of weed control, which are suppressed by fast-growing sideral plants and do not have time to inseminate when harvesting the siderate in the flowering phase. Yield increase when using siderate reaches 1.6 t/ha, and costs are 4-5 times less than when transporting and applying organic fertilisers.

The main tillage for beetroot is considered to be particularly energy-intensive. It is recommended to cultivate clean fallow by the early fallow type, starting with post-harvest no-tillage loosening to a depth of 10-12 cm in autumn. In early summer, mouldboard ploughing to a depth of 25-27 cm, then shallow cultivation in combination with chemical weeding in the appearance of weeds. At the end of autumn - no-till deep loosening under beetroot at a depth of 28-30 cm. According to a number of researchers, it is possible to replace deep mouldboard ploughing with shallower ploughing (20 cm) and combine it with deep no-tillage loosening without noticeable reduction in yield. The less costly no-tillage basic cultivation for beet in dry conditions is not excluded, but weeds, especially perennial weeds, are more widespread and herbicides should be used more widely.

The field is then harrowed and levelled with harrows in early spring. But if the soil has been thoroughly levelled since autumn by post-ploughing cultivations, then in areas where the earliest sowing is practised (e.g. in the steppe on 1-5 May), early spring cultivation can be avoided and sowing can be done immediately after pre-sowing tillage. Pre-sowing tillage should be to a depth of no more than 3- 4 cm in order to create a firm seedbed. It is important not to disturb the capillary system formed in the soil during winter and spring, through which moisture comes to the seeds and roots. Therefore, it is better to do the treatment on the day of sowing, using USMK-5,4 or combined aggregates combining loosening, levelling and rolling.
Beetroot sowing should be started when the soil at a depth of 5-7 cm warms up to +6...+80C. In the steppe it is the first decade of May, in the forest-steppe - the beginning of the second decade. Late sowing leads to loss of moisture from the topsoil, shortening of the vegetation period and, as a result, to poor yield and lower sugar content. Each day of delayed sowing leads to a 3-4 centners/ha reduction in yield.
The most important technological method is sowing and formation of optimal plant density. Practice shows that the actual density of beet plants is often 60-70% of the recommended, which entails corresponding yield losses.
Modern European intensive technologies provide for precision sowing to a final density of 6-7 fruits per 1 m of row, which is 1.3-1.4 seed units per 1 ha. One sowing unit is the number of seeds sown per 1 ha when sowing 5 fruits per 1 m of row. With precision sowing - the minimum consumption of expensive seeds, and in the future there is no need for operations to form the density of plants. But precision sowing is possible on fields clean from weeds when providing effective means of plant protection against diseases, pests, weeds during vegetation, as well as in the presence of high-quality seed with germination of at least 90-95%, equalised by 90%. At the same time, it is necessary to create conditions for high field germination.
Experience shows that in the Altai Territory field germination of beet seeds is 40-60%, so the final density is sown 10 pieces per 1 m of row (seed consumption of 3.5-4.5 kg / ha) to get 5-6 pieces of sprouts, which by harvesting will be 4-5 pieces per 1 m of row, that is, about 100 thousand plants per 1 ha. This is the optimal density of beet plants in the forest-steppe. In steppe, the optimal plant density is 80 thousand plants per 1 ha, or 3-4 plants per 1 m of row.
If there are limited possibilities to ensure plant protection, the sowing rate should be increased to 15 fruits per 1 m of row and, depending on the seedlings obtained, thinning should be done if necessary. This

sowing option is possible only if there is sufficiently cheap seed material. But at present, in the absence of sugar beet seed production in Altai Krai, seeds from outside the region are used, the price for which is very high, so mostly only precision sowing is practised. In any case, the field should be sown in 1-2 days.

The sowing depth is 2.5-3 cm. Beetroot does not tolerate deep sowing of seeds, as the weight of the seeds itself is 4 g (with the weight of 1000 fruits being 15 g), the reserve of nutrients in the seeds is 10 times less than that of wheat. The sprout brings the seed pods to the surface and is not protected as in cereal crops by the colioptile. Therefore, beetroot should be sown shallowly, but in a moistened soil layer on a dense bed. The sowing speed should be 4.5-5 km/hour. Dry soil should be rolled before and after sowing.

Pneumatic seed drills, both foreign (OPTIMA) and domestic (STV-12), as well as mechanical seed drills SST-12B when equipped with improved seeding discs, provide precise seeding. The importance of seeding accuracy increases in intensive technologies.

The seed is of great importance. Hybrids are the basis of intensive technologies. Foreign hybrids surpass domestic hybrids in terms of yield, as a rule, by 2025%. Their advantages are, in particular, in better seed processing, such as high-quality draping. Dredging is giving seeds a spherical shape by applying to their surface a mixture that includes a dressing agent, dusty peat, organic and mineral fertilisers, growth stimulants, microelements and adhesives. The seeds are also ground, that is, the fruit coat is partially removed, after which the seeds become more friable. Inoculated seeds treated with a dressing agent in combination with an adhesive are also available. Seed preparation is carried out in seed factories.

Dredged seeds should be sown in a well moistened soil layer. Under conditions of lack of moisture, no more than 30 per cent of the sown area can be sown with drained seeds.

There is evidence that foreign hybrids are more susceptible to diseases both during the growing season and during storage before processing, as European technologies provide for more disease control by chemical methods, and beetroot coming for processing is quickly processed. The cost of imported seeds is an order of magnitude higher than domestic seeds, and the expected high additional gains can be obtained only under conditions close to optimal.

If the choice is between domestic hybrids, undoubtedly, preference should be given to the released ones, as they have confirmed their high productivity in the conditions of Altai Krai. In conditions of extensive technology, designed for lower yields, the importance of varieties-populations or hybrid varieties as more plastic, easy to seed and less expensive increases.

Modern intensive precision seeding technologies do not require pre-emergence and post-emergence tillage to prevent beet thinning, but herbicides are used to suppress weeds. Both soil and vegetative herbicides or combinations of both are used. In conditions of unstable and insufficient moisture a higher effect is given by herbicide application under pre-sowing cultivation at a depth of 3-5 cm in the soil layer where the main mass of weeds is located. For application, a boom sprayer is used, aggregated with a cultivator USMK-5,4 with ring-spur rollers.

According to many experiments, the relative yield increases up to 120-130% with effective use of herbicides. The yield gain of 30 to 45 kg/ha is close in value to the cost of chemical defence.

For farms that cannot afford the use of expensive herbicides during the growing season, it is possible to apply technologies that provide for weed control to a greater extent through mechanical methods: more intensive tillage before sowing, pre- and post-emergence harrowing, inter-row cultivation until the leaves close in the inter-row (the latter is done with beet perching in the rows), mowing of weeds at the height of beet plants. Plant thinning is prevented by a higher seeding rate, and harrowing is used as one of the methods of forming optimal plant density.

When the topsoil is dry, rollers are rolled at a speed of 6-7 km/hour.

On the 4th-5th day after sowing, pre-emergence soil loosening is carried out to destroy soil crust at a depth of 2/3 of the sowing depth. This can be done with light seed harrows (ZBP-0,6) across the sowing or cultivators USMK-5,4 with or without rotary working bodies and bar bodies, along the rows according to the tractor's track marker. If beet sprouts reach 0.5 cm, it is dangerous to carry out pre-emergence harrowing.

At the emergence of beet sprouts, beetroot is carried out spheres - shallow loosening of the soil in the inter-rows to a depth of 3-4 cm, using USMK-5,4 with ploskreskozuyuschie razors and rotary batteries, as well as protective discs on each side of the row to protect the seedlings from being covered.

After the first inter-row cultivation in 3-4 days (when beetroot has the 2nd pair of true leaves) and the presence of more than 6-7 plants on 1 m of row, harrowing of seedlings with harrows ZBP-0.6 or rayboronki ZOR-07 at a speed of 3-4 km/hour, which will improve soil aeration and destroy weeds in the row zone. It is better to do it in the daytime, as beetroot is strongly damaged at high turgor.

The next inter-row cultivation at 2-3 pairs of true beet leaves is carried out without protective discs using the same loosening tools as in spheroidal cultivation. The depth of loosening with a chisel in the centre of the row spacing is 7-8 cm. To cover the weeds in the rows use perching tools with a loosening depth of 4-6 cm. Weeds in the rows are covered with a 1 cm layer of soil, except for the growth points of beet plants. Before closing the rows, inter-row cultivation is also carried out with perching. Cultivation is used as an additional method of weed control in the protective zone, where the cultivator does not loosen the soil. In addition, covering of beet plants contributes to their more uniform development and, therefore, better harvesting, as it reduces damage to root crops and clogging of the pile with earth during digging, provides more stable operation of automatic row driving devices.

The most common pests of sugar beet are wireworms, beet fleas and grey beet weevil.

Wireworms accumulate in the fields of crop rotation, especially if there are weeds, so it is necessary to clear the fields of weeds. Wireworms (larvae of click beetles) damage sugar beet seeds, their seedlings and plants, especially in the initial period of development.

Beet beet flea beetles cause damage during the period of seedling emergence up to the phase of two pairs of true leaves. In hot, dry weather during the fork phase, if beet seeds are not treated, beet flea can completely destroy beet seedlings within a day or two.

The grey beet weevil, like the wireworm, accumulates in fields, especially where there is a large number of perennial weeds. The beetles damage seedpods and young leaves.

The problem of controlling wireworm, beet flea and grey weevil is solved by treating seeds at the seed factory with systemic seed dressing agents, which are effective not only against diseases but also against pests. If the seeds are not treated, it is necessary to prepare in advance for pest protection of beet seedlings by using insecticides.

Sugar beet is affected by diseases, the most common of which is the rootworm. It affects seedlings up to the phase of two or three pairs of leaves. The disease increases sharply if heavy rains have fallen after sowing before sprouting and a soil crust has formed, as well as if the beet is not properly positioned in the crop rotation and if there is a lack of nutrients. During the first period of beet root damage (especially on cloudy days), when inspecting the above-ground part of the plants, the disease is not noticeable, but when hot and dry days arrive, the death of seedlings reaches 30-40%.

Measures to combat root-knot disease. Beet in crop rotation should be returned not earlier than in 4-5 years, organic fertilisers should be applied to beet in combination with mineral fertilisers at the optimum ratio of nutrition elements, sowing should be carried out at the optimum depth at an average daily air temperature of 13-16^{0}C., good aeration of the soil from sowing to 3 pairs of true leaves by pre-sowing and post-sowing loosening, sowing should be carried out with quality seeds, provide good soil aeration from sowing to 3 pairs of true leaves by means of pre-and post-emergence and post-emergence loosening, sowing with quality seeds with germination of not less than 85-90%, treated with highly effective systemic dressing agents.

Harvesting. The optimal harvesting period in the Altai Territory is from 15 September to 10 October, as harvesting earlier leads to a lack of sugar in root crops, and harvesting after 10 October provokes the risk of leaving the crop in the field undigested.

Before digging, the haulm is mown using a BM-6 haulm harvester. Harvesting costs account for about 40% of the total costs of cultivation. Yield losses reach 30-40% due to imperfect harvesting equipment and inadequate technology. In particular, when harvesting the haulm, the cutterbar is set to a certain cutting height. With uneven placement in the row, the beet plants form root crops of different sizes. In the sparse state, the larger root crops protrude more above the soil surface, while in the smaller root crops are more deeply embedded in the soil. It is therefore inevitable that a certain proportion of the larger root crops will be lost when the haulm is cut low. In any case, it is necessary to strive for an even distribution of plants in the row.

According to VNIISS data, application of improved harvesting technology, when BM-6B at high cutting height removes haulm, but at the same time works with head after-cleaner OGD-6A, allows to reduce losses to a minimum with quality haulm removal in comparison with haulm harvesting BM-6B at medium cutting height without OGD-6A.

Harvesting methods. The flow method is used on lighter and drier soils, when the total contamination of root crops is not more than 10%, including haulm not more than 3%, as well as when there is enough transport in the farm and when transported at a distance of not more than 15 km. During in-line harvesting, root crops are dug by combines KS-6, RKS-6, "Hol-mer", loaded and transported to the beet harvesting station.

In case of increased contamination on heavier and wetter soils, the transshipment method is used, when the excavated root crops are unloaded into kagats at the edge of the field (on a previously prepared and cleaned area), if necessary, cleaned from the haulm, loaded into transport with the help of beet loader SPS-4,2 and taken to the beet point. The need for transport is reduced by 50-60%, with double loading and unloading. The transshipment method of harvesting consists in the fact that along with transport from the harvester, a part of root crops is unloaded into cages.

Optimal terms and modes of beet harvesting are possible at seasonal load on domestic machines BM-6, KS-6, RKM-6 of 80-100 ha, and on imported harvesters combining harvesting of haulm and digging of root crops simultaneously - HOLMER, KLAINE, VKM-9000 - up to 500-600 ha.

Requirements for the quality of harvesting: cutting haulm should be made without chipping heads and leave on it no more than 3% of petioles, waste sugar-bearing mass - no more than 5%, the number of root crops knocked out of the soil - no more than 0.5, losses of root crops and their parts in the soil - no more than 3, total contamination - no more than 10, including haulm - no more than 3%.

Requirements for root crops delivered to the sugar factory: green mass - not more than 3%, the number of injured root crops (lost more than 30% of the volume) - not more than 12, flowering root crops - not more than 1, wilted - not more than 5%.

Frozen or substandard beetroot for any of the above reasons is accepted as substandard beetroot with a 20% discount in price. A premium is paid for each percentage of sugar content above the basic sugar content. Base sugar content is the average sugar content in the zone over the last 5 years.

Peculiarities of fodder root crops cultivation. It is recommended to place fodder root crops in farming or fodder rotations in order not to transport the crop and organic fertilisers over long distances. They should not be placed after related crops, e.g. rutabaga and turnips after mustard and rape. Optimal predecessors are winter cereals, annual fodder grasses (pea-pea-oat mixtures), melilot. Cultivation technology is similar to the technology for sugar beet. The following features can be noted: fodder root crops are sown in early terms, carrots can be sown in winter terms, turnips - from the end of May to the middle of June, i.e. in summer terms, as early terms make turnip root crops worse stored. Vegetable seeders SON-4,2, SON-2,8 can be used for sowing. The sowing method is wide-row sowing - 45 or 60 cm, carrots can be sown by ribbon sowing. Sowing depth is 1.5-2.5 cm. Seeding rate: carrots - 1.5 million/ha - when sown in a wide row, 2-3 million/ha (4-5 kg/ha) - when sown in a ribbon method, turnips and rutabaga - 0.5-0.8 million/ha (1.52.5 kg/ha). For more uniform sowing of small-seeded crops use sowing together with ballast, for example, with superphosphate, sieved through a sieve with holes of 2-4 mm, with a consumption of 20-25 kg / ha. Recommended density in cultivation: fodder beet - 65-80 thousand/ha, rutabaga - 50-90, turnips - 80100, carrots - 300-350 thousand/ha.

Before harvesting, the haulm is mown with BM-6, KIR-1.5. The haulm can be fed to animals, silage, make grass meal. To dig up root crops use potato diggers UKV-2, converted potato harvesters KKU-2A, forage root digger KKG-1.4. It is better to harvest in a flow method and immediately take it to the place of storage. Forage root crops are stored in standard storage facilities, in concrete trenches, in piles covered with earth and straw. Storage temperature +1...+2^{0}C, relative air humidity 85-95%. Turnips are stored somewhat worse than rutabaga, they are fed first.

10.5. Cultivation of mother beet and seed beets

Root crops grown in the first year of life and stored for seed production the following year are called mother beets. Mother beet is sown with elite seeds grown in elite seed farms. Elite seeds are delivered to seed farms, where the mother beet is grown in the first year, and in the second year planted mother root crops, get a crop of seeds of the 1st reproduction. These seeds are sown in sugar beet farms, where they produce a crop of factory beet for processing at sugar factories.

Peculiarities of cultivation of mother beet. Placement is carried out on the same predecessors as in the case of factory beet, but with observance of spatial isolation of at least 1 km from planting and last year's beet fields to prevent infection. Baby beet is sown a little later than factory beet (late May - early June), grown at a higher density: 160-180 thousand/ha in areas with sufficient moisture, 120-140 thousand/ha - in areas of insufficient moisture, leaving 10 plants per 1 m of row. Under such conditions, plants form a small root crop (150-300 g), suitable for mechanised planting. Increases the yield of mother plants from 1 ha. Smaller root crops are better stored due to the formation of small cell structure of xerophytic type, and also have higher seed production the following year.

Do not allow root crops to freeze or dry out during harvesting. Frozen root crops rot in storage, and wilted root crops give a large number of "stubborn" - non-flowering plants - the next year. In Altai, harvesting of mother plants should be carried out from 25 September to 1 October, when the average daily air temperature decreases to +6...+80C, but there are no frosts yet. This reduces plant respiration and the risk of diseases, as microbiological activity is reduced. Before harvesting, diseased plants with signs of peronosporosis, rot, mosaic, blossom blight and fodder beet should be removed. Cutting of haulm (BM-6) is done with leaving the petioles at a height of 3-4 cm, so as not to damage the buds of renewal on the head. Stored mother root crops in storages or cages at a temperature of +2...+30C. In spring, healthy root crops with a diameter of 5-10 cm and cone-shaped are selected.

Cultivation of plantings. Beetroot of the second year of life is more water-loving, as it has a weaker root system and high transpiration coefficient (K_{tr} = 725), so the planting is planted only in pairs. The planting date is the earliest - in early May. Before planting make cultivation to a depth of 16-18 cm. Use for planting planting machine VPU-4 or VPS-2,8. Planting scheme - 70x70 cm with a density of 20.4 thousand/ha. It is possible for smaller root crops scheme 70x35 cm. The head at planting should be submerged to a depth of 2-3 cm, and the soil should tightly encircle the root crop. When 30% of sprouts appear, harrowing is done, and during vegetation - at least 3 inter-row cultivations.

Isolation between plantings of different varieties - 1 km, diploids and tetraploids - 3, single-seeded and multiseed - 5, between different subspecies - 10 km.

Special plant care techniques: cutting off the central bud to form a multi-stemmed plant, seedpod pincing - removing the tops of flower-bearing shoots at the beginning of flowering by 2-3 cm so that the plants do not outgrow, additional pollination with a rope - 3 times in 5 days.

Harvesting of seedlings is started when 30-40% of fruits on the plants turn brown and seeds are powdery on the break. Grain harvesters with reapers ZhBA-3,5, ZhRS-4,9 mow 5-7 rows into a swath. After ripening, combine harvesters equipped with canvas-plank pickers PTP-2,4B, PPT-3,0A, PTP-3,0 are used for threshing. Immediately after harvesting it is obligatory to clean seeds on OVP-20A, then on beet hill or on SM-4, ZAV-20. Seeds are stored at humidity not more than 15%.

Plantingless seed production of sugar beet consists in the fact that in areas with mild climate (Stavropol, Krasnodar Territory, Kyrgyzstan) the sugar beet is not dug up, it overwinters in the field, the next year it grows and gives a crop of seeds. For better overwintering, the plants are covered with earth with a layer of 10-15 cm by perching. In spring they are unhooked and inter-row cultivation is done. Seed yield is somewhat lower, but the cost of production and labour inputs are 2-3 times less.

11. CARTOFEL

11.1. Importance, yield, distribution

Potatoes are an important food crop. Potato tubers contain 25% of dry matter, including starch - 14-22, protein - 1.4-3, fibre - 1, fat - 0.2-0.3, ash - 1%, vitamins C, B_1, B_2, B_b, PP, carotenoids. Potatoes are used to produce starch, molasses, alcohol, glucose, rubber and other products. From 1 tonne of tubers 112 litres of alcohol are obtained. Potatoes are used as fodder for cattle. The digestibility of tubers is 83-97%. 100 kg of raw tubers contain 29.5 k.u., silage from haulm - 8.5, fresh barda - 4, fresh pulp - 13 k.u. Bard is a waste product of the alcohol industry, pulp is a waste product of starch production. The peel of greenish tubers contains alkaloids solanine and chaconine, which decompose during cooking.

The agronomic importance of potato is that it is a good precursor for wheat and legumes, as it leaves the fields clean from weeds. Early potato varieties can be used as a vapour-occupying and insurance crop.

The potato is very plastic, widely distributed throughout the world and covers an area of about 18 million ha with a gross production of 265 million tonnes and a yield of 14.6 tonnes/ha. In Russia, the area under potato cultivation is about 3.3 million ha, mainly in the Central and Central Zones, the Volga region, Bashkiria, the Urals, Siberia and the Far East. The share of the public sector in potato production in our country has decreased to 10 per cent, the rest is obtained in the private sector. The average yield in the country is 10-11 tonnes/ha, although in some farms with intensive technology it reaches 25-30 tonnes/ha, including at the West Siberian vegetable and potato station. In European countries, the yield is 30 tonnes/ha and higher.

11.2. Botanical characteristics of potatoes

Potatoes belong to the Solanum genus of the Solanaceae family. The cultivated potato is Solanum tuberosum. In Central and South America, wild species Solanum demissum, S. Andigenum are also widespread, which are used for the cultivation of potatoes

in breeding as resistant to diseases and pests. Potato is a perennial herbaceous plant, but in culture it is used as an annual. The main method of potato propagation is vegetative, by means of tubers. A tuber is the thickened end of an underground stem shoot (stolon). The root system of potatoes grown from a tuber is lobe-like, but when growing from seed, a sprout with two seedpods and an embryonic stem root is formed first, then adventitious roots appear on the underground part of the stem and on the stolons, just as when growing from tubers. Potato roots penetrate shallowly, located mainly in the arable horizon, and only some go to a depth of 1 -1.5 m and to the side for 0.5 m. The root system of potatoes has a fairly high absorption capacity, especially in relation to phosphorus, but weakly overcomes the mechanical resistance of the soil.

The tuber has spiralling eyes, each containing up to three buds. There is a small scale above the eye (a modified reduced leaf), and after it shrinks and falls off, a small scar remains. There are more eyes at the top of the tuber and fewer at the base, where there is a stolon depression. Tubers can be of different shapes (rounded, elongated, oval), with white, yellow and blue flesh colouring. Yellow-meat varieties are richer in carotene. Mature tubers are covered with a cork tissue skin, which protects them from drying out and diseases. Coloured tubers are pink, light red, light blue, etc. The colour of the tuber coincides with the colour of sprouts and flowers.

When a tuber is planted, shoots form from the buds, with roots and stolons forming in the underground nodes. One plant produces up to 6-8 underground stems and stolons, each of which can branch. Stolons grow up to 25-30 cm long, then thicken at the ends to form new tubers.

The potato leaf is compound, interruptedly-parietal-peristocratic-sectioned, consists of lobes, lobes and lobes, is weakly, moderately and strongly dissected, pubescent.

The inflorescence consists of 2-4 whorls on a long peduncle. The flower has 5 (less often 6) petals, partially fused, white or coloured (pink, red-violet, blue-violet), 5 stamens and a pistil. The anthers may be yellow, orange, green. The potato is self-pollinating. It often has widespread male sterility due to pollen sterility, so it often does not produce fruit or seeds, but this does not affect tuber yield. The fruit of the potato is a juicy two-nested berry. The weight of 1000 seeds is 0.5 g.

11.3. Biological characterisation

Potato is a rather heat-loving plant. In mature tubers buds start to grow at +3...+50C without root formation, and with root formation - at 17...112^{0}C. The optimum temperature for the sprouting period is +15^{0}C, then the sprouts appear in 15 days, at low temperatures - in 20-25 days. A temperature of about +20^{0}C in the soil is optimal for the growth of haulm and tuber formation. High temperatures above +25^{0}C cause increased branching of stolons to the detriment of club-formation, and +30^{0}C and above - paralyses assimilation and growth.

Freezing -2^{0}C (5-6 hours) causes death of young plants, blackening of haulm. At a temperature of 0...+10C in tubers starch turns into sugar, tubers become sweet, but if they are not frozen, they can be kept for 5-10 days at room temperature, the sweet flavour disappears, as sucrose is spent on more intensive respiration of tubers at high temperature. Tubers out of the ground at -1.2^{0}C freeze as they are 75% water. Frozen tubers rot during storage.

The vegetation period of potatoes is from 60 to 180 days, the sum of active temperatures for the vegetation period depending on the variety is 1000-1600^{0}C.

Potato is a moisture-loving plant, as it has a weak root system. The yielding part (tubers) consists of 75% water, so if there is a lack of moisture during the period of tuber formation, the yield is greatly reduced. During the germination - sprouting period, the plant takes water from the mother tuber. The critical period in relation to moisture is budding - tuber formation, because from the budding phase on the ends of stolons begins tuber formation, and in this period their number is determined. The optimum moisture content is 70-80% of NV. During the period of starch accumulation, the optimum moisture content is 60% of NV. The transpiration coefficient can be from 160 to 650, which indicates the high plasticity of this crop. Potatoes can absorb moisture partly from the atmosphere through the leaf surface.

Potatoes have high requirements for air nutrition, as it consumes a lot of oxygen through the root system (up to 1 mg per 1 g of dry matter, which is 5 times more than sunflower), so the soil must be constantly loose, otherwise tubers suffocate and rot, as evidenced by the appearance of loose white lenticles.

Potatoes are a crop of loose, sufficiently fertile soils with humus content of at least 2%. The most suitable are sandy loam and light loamy chernozems with a volumetric weight of 1-1.2 g/m^3. On light sandy soils tubers are formed with high flavour qualities, but potatoes suffer more from lack of moisture and potassium on such soils. Cultivated peatlands are good for potatoes, especially for seed potatoes. Potatoes can be grown on fertilised sandy and sod-podzolic soils and grey forest soils at . Heavy, clayey, strongly compacted soils with close groundwater table are unsuitable. On such soils tubers are formed small, deformed, instead of sprouts can be formed young tubers. Saline soils are unsuitable for potatoes, but they tolerate slightly acidic soils well. The optimum pH level is 5-6.

Potato is a short-day plant by origin, but in mid-latitude conditions it tolerates a long day well. As it moves southwards under short day conditions, it accelerates its development, shortens the growing season, including the duration of tuber formation and growth, so tuber yields in the south tend to be small. This is a light-loving crop, and if there is a lack of light, the stems elongate, leaves turn yellow, flowering is absent, and yields decrease. If the rows are orientated from south to north and have good light, the yield is 20% higher.

The following periods are noted in the proccss of potato growth and development.

1. The dormancy period is divided into natural dormancy, when tubers do not sprout after harvesting, even under optimal conditions for 2-3 months, and forced dormancy, when potatoes are stored at reduced temperatures of +2^{0}C to prevent sprouting. Tubers can be brought out of natural dormancy by keeping them at 0^{0}C ^{for} 2-3 days. This may be of value in double-harvest potato culture in southern areas.
2. Sprout establishment period (from planting to sprouting), when the sprouts, root system and underground part of the stem are formed. It can last up to 20-25 days, so blind cultivation during this time is important.
3. Sprouts - budding. At this time stolons grow, the main assimilating surface and roots are formed.
4. Boutonisation - flowering. Stolons stop growing in length, thicken at the ends, tuber formation takes place, the number of tubers in the nest is formed.
5. Tuber growth. Assimilation products enter the tubers.
6. Withering of haulm, coarsening of tuber covering tissues, physiological ripeness.

Different groups of varieties are distinguished by the length of the growing season (Table 33).

Table 33

Potato variety groups by maturity

Variety groups	Days from planting	
	before the formation of marketable tubers	before the haulm starts to die
The early ones	55-65	80-90
Medium-early	65-80	100-115
Mid-maturing	80-100	115-125
Mid-late	100-110	125-140
Late maturing	More than 110	More than 140

In farms it is more favourable to have 2-3 varieties of different maturity, which will allow to obtain more stable yields in different weather conditions in years, as early varieties use precipitation in June more effectively, mid-maturing varieties - in July, mid-late varieties - in August. The same will allow to carry out all works in optimal terms and reduce the need for machinery and labour. Early maturing varieties may be slightly lower yielding and have lower starch content (14%), but they produce the earliest, freshest produce in mid-summer. Late maturing varieties with high starch content (20-22%) are used as technical varieties for processing. Table varieties have a starch content of 1419%. Starch content slightly decreases (by 1 -2%) with increasing doses of fertilisers and irrigation.

11.4. Food potato cultivation technology

Place in crop rotation. With good tillage and application of fertilisers, potatoes tolerate repeated planting well. Its specific weight in the crop rotation can be from 25 to 50%. In a crop rotation with repeated planting for phytosanitary purposes it is necessary to provide for the use of intercrops, sidedresses, perennial grasses, organic fertilisers. For example, the crop rotation can be as follows: 1) early potatoes in fallow; 2) winter rye; 3) potatoes; 4) oats +

+ clover; 5) clover; 6) potatoes; 7) potatoes; 8) vetch-oats. The best precursors for potatoes are winter and spring cereals, legumes, turnover of perennial grasses, fallow. In vegetable rotations potatoes should not be placed after nightshades (tomatoes, peppers, aubergines). Early potatoes can be used in occupied fallow or in crops after winter rye for green mass. The compaction crop is relevant in suburban farms where there is little arable land.

Mineral nutrition and fertilisation of potatoes. Due to the natural fertility of the soil, a yield of 815 tonnes/ha can be expected. With 1 tonne of tubers and the corresponding amount of haulm potatoes take out N 5-6 kg, P2O5 - 1.5-2, K2O - 7-10 kg, that is much more than cereals. Potassium yield is 4 times more than that of wheat. Potatoes are considered a potassium-loving crop. Potassium increases resistance to diseases, improves the storability of tubers, protein and carbohydrate metabolism, tuber formation, increases starchiness along with phosphorus. Nitrogen and phosphorus are equally important, the lack of which reduces the productivity of photosynthesis, slows down plant development.

Organic fertilisers are of great importance for potatoes. The rate of organic fertiliser application is 20-40 t/ha, and on poor soils - up to 40-60 t/ha. 1 tonne of organic fertiliser gives a yield increase of 0.2 tonnes/ha. Peat-dung composts are very effective: 20 t of manure + 20 t of peat + 10 t of sod + 5-6 t of potash fertiliser or wood ash.

If organic matter is applied under the predecessor, only mineral fertilisers are applied to potatoes. Possible application rates on chernozems are N45-60P60-90K30-45. On different soils, in different conditions, the norms of fertiliser application are specified. It is better to apply chlorine-free potassium fertilisers to potatoes, as chlorine deteriorates taste qualities and reduces starch content, or apply KC1 from autumn.

Microfertilisers copper, boron, molybdenum, zinc increase resistance to diseases, starch content, reduce nitrate content in tubers. Before planting tubers are treated in 0.05% solution of appropriate salts or sprayed with the same solutions in the budding phase.

Tillage. Potatoes are very responsive to deep mouldboard ploughing. After harvesting the stubble predecessor stubble husking is carried out, 2-3 weeks later - mouldboard ploughing to the depth of the arable horizon. In erosion-prone areas, no-tillage is done. In spring - RWB in 2 traces, before planting - cultivation to a depth of 12-14 cm. Such a complex is used on light soils. On heavier soils in the spring do ploughing to a depth of 22-27 cm ploughs without mouldboards, but with skimmers set at 12-14 cm, or stands SibIME with harrowing. This ensures loosening of arable horizon without extraction of soil blocks and clods. The use of plough cutter for spring tillage for potatoes is promising, which provides very good crumbling of soil and mixing it with fertilisers, which is especially effective on heavy soils.
Preparation of tubers for planting. Sorting or calibration of tubers into fractions: 25-50 g, 50-80, 80 g and more with the help of KSP-25 (potato sorting station). The 50-80 g fraction is used for planting . If there is a shortage of planting material, large tubers are cut along the tuber so that each half had 2-3 eyes. Cut tubers lose their protection in the form of cork tissue, so it is better to wilt them and treat them with pesticides. Using large tubers for planting can increase yields, as they have more buds, they give earlier sprouts, but this does not justify the high cost of planting material. In addition, large tubers are poorly stored.
10-12 days before planting tubers are picked, remove diseased.
Germination of potatoes before planting for 20-30 days at a temperature of 12-15^{0}C in the light is very effective. In the dark, potatoes sprout long, thin etiolated sprouts that break off when planted. In the light, potatoes form short but strong thickened sprouts, which should be no more than 1 cm. Sprouted potatoes provide faster development, which is especially important for early potatoes and in areas with a short growing season. The yield increase when planting sprouted potatoes is 7 to 10 tonnes/ha. Wilting for 6 days at +25^{0}C or 2-3 hours at +35-40^{0}C is also effective. This accelerates enzyme activity and partial greening of tubers.
For protection against diseases, tubers are treated using, for example, Maxim KE, 0.2-0.4 litres/tuber. It is possible to combine dressing with microelements. Soaking tubers in a solution of mineral or humic fertilisers is also used.
Planting. Potatoes can be planted when the soil at a depth of 10 cm warms up to +70C. In Altai, this is the middle of May. Earlier planting (at +5^{0}C) is possible for early-ripening more cold-resistant varieties, as well as when planting sprouted tubers. The order of planting varieties of different maturity should be as follows: first plant early maturing

potatoes to get the earliest possible production, then late maturing and medium maturing varieties.

The planting depth of potatoes is 6 to 15 cm. A shallower planting depth of 6-10 cm is recommended for early potatoes, for planting smaller tubers and for cultivation on heavier soils. In areas with less moisture and lighter soils, potatoes respond better to deeper planting. Potatoes should be planted deeper for seed purposes, with this a multituber nest of relatively small and physiologically younger tubers is formed, the yield of seed fraction increases, the yield of healthier and more productive plants increases, as sick tubers from a greater depth do not sprout.

Potatoes are planted in wide rows with row spacing of 70 cm and a row spacing of 25-35 cm. There are different planting methods, which have their advantages under certain conditions. Smooth planting is recommended when there is a lack of moisture, in the steppe zone and on lighter sandy and sandy loam soils. The arable surface remains smooth until harvesting.

In more water-secure areas in the forest-steppe, on heavier soils, ridge planting is effective. Ridge cutting can be done in autumn, in spring before planting, in time or after planting with cultivators KRN-4.2 with perching devices. The advantage of autumn ridge cutting is that in winter under the action of frosts soil clods are destroyed, the soil in the ridges warms up 7-10 days earlier, soil temperature - 34^{0}C higher than on a flat surface. This allows earlier planting and more efficient use of moisture reserves. In ridges potato increases the volume of root layer, increases aeration, and thus improves air and nutrient regime, additional stolons and tubers are formed, and the yield increases. There is a possibility of more effective band fertiliser application. During harvesting, losses are reduced because the soil is more loose, the load on the machine's digging organs is reduced, and productivity is increased.

Ridge-and-tape planting is also used in areas with excessive moisture, such as in the Far East.

Potatoes are planted in beds 35 cm high and 140 cm wide at the base. The planting pattern is (110+30) x 30 cm. The potatoes are arranged in close rows with a distance of 30 cm between them. The excess moisture drains to the bottom of the bed and the tubers rot less. This planting has all the same advantages as row planting. This planting technique is also known as the belt-row planting, and it produces higher yields not only in areas with excessive moisture, but also in areas where potatoes are traditionally grown. The root system of plants is less damaged in ridges, as soil compaction by tractor wheels during inter-row cultivation does not reach the tuber nest zone. During harvesting, 30-40% less soil is delivered to the combine's conveyor belt from the ridges than with ridged planting.

For planting use potato planters SN-4B, KSM-4A, KSM- 6A, L-201, L-202 for planting in ridges, SKM-3A - for planting in ridges, SAYA-4 - for planting with sprouted tubers. Planting speed is 4.5-6 km/hour.

Planting density - 55-60 thousand/ha, in less water-saturated areas - 45-50 thousand/ha. Early varieties with a more compact bush are planted at a density of 65-70 thousand/ha. In dense plantings potatoes accelerate development, increase starch content, which is important for early potatoes. The weight planting rate is about 3 tonnes/ha.

Potato planting care starts very early. The pre-emergence period is very long, many weeds sprout, which are destroyed by pre-emergence harrowing twice with an interval of 5-7 days. When ridged planting can do inter-row cultivation even before sprouting cultivators KON- 2,8, KRN-4,2, and in this case in the row zone set net harrows. After sprouting, one to three inter-row cultivations are made depending on soil density and weediness. It is recommended to keep the depth of cultivation at 10-12 cm - in the first treatment, 6-8 cm - in the second. Protective zone in the inter-row treatment increases from 10 cm on each side of the row - in the first and up to 15 cm - in the second treatment. Inter-row cultivation is combined with perching (KPH-4,2 with perchers). The effect of perching is the same as with ridge planting, so the earlier perching is carried out, the more effective it is, but at the same time the soil should have enough moisture and nutrients. In a dry steppe it is better to avoid perching, as loosening causes moisture loss.

Up to 30% of potato yield is lost due to diseases and pests. From Colorado potato beetle it is recommended to spray plants with solutions of Karate, Decis 0.2 l/ha 2-3 times, the first time - at mass appearance of larvae, the second - after 10-12 days. Biological preparations for Colorado potato beetle control are used in biological crop production: Bitoxybacilin, Colorado, Bicol based on VasP- lus thuringiensis.

Wireworm control measures include weed control, especially of wheatgrass, winter ploughing, and application of ammonia fertilisers. Potatoes grown after peas are less damaged by wireworms. Mustard and rape, used as intercrops, clean the soil from nematodes.

Phytophthora is the most dangerous among potato diseases, especially in water-saturated areas. This disease manifests itself more in case of excessive nitrogen fertilisers and lack of microelements. Control measures : observance of crop rotation, dressing of planting material with cuprozan, cineb (1.5 kg/t), mowing of haulm in advance before harvesting, drying and light hardening of tubers before storing, landscaping, spraying of plants with systemic preparations starting from the budding phase, spraying of plants with copper preparations (1% solution of copper sulfate or Bordeaux liquid).

Common potato scab is a fungal disease that affects potatoes, especially on light soils in hot, dry weather. The same disease can be caused by fresh, unlittered manure.

Bacterial diseases - ring rot, black leg, wet rot. For recovery from pathogens of diseases good predecessors - winter rye, perennial grasses, alkaloid lupine. Drying, harvesting, greening of seed potatoes are obligatory.

Viral diseases - mosaic, jaundice. Signs - shriveled and yellow leaves. Gothic causes small narrow leaves with anthocyanin, spindle-shaped tubers, chlorosis, necrosis, defoliation. Control measures: resistant varieties, crop rotation, culture of apical meristem in seed production system, sowing of white mustard along the field perimeter to control virus vectors, phytocleaning.

Potato harvesting costs up to 60 per cent of the total costs. Potatoes are physiologically mature when the haulm has wilted, the tubers have a thick, flaky skin, and the stolons are dry and easily separated from the tubers. 2-7 days before harvesting mow the haulm to dry the ridges, rapid ripening, prevent infection, improve the quality of tubers. Desiccation 10 days before harvest by spraying with magnesium chlorate (25 kg/ha) is effective on diseased and late maturing varieties. The haulm is mowed with KIR-1.5.

Potatoes on light and medium soils with humidity not more than 25% are harvested in a flow method by potato harvesters KKU-2A, KPK-3, E-686 and others at a speed of 3 km/hour. The dug potatoes are immediately loaded into transport.

Separate harvesting is used on medium and heavy moist soils. Potatoes are dug out with a windrower VKV-2, capturing two rows in one pass and placing them in a windrow for drying. At low yields in the same windrow is laid potatoes from two or four more rows in the next pass of the same machine. Then KKU-2 picks up potatoes from the swaths at higher speed.

Combined harvesting is used in fields with low yields and reduces labour costs by 30%. The VKV-2 windrower places tubers between two adjacent unploughed rows. Then the KKU-2A harvester digs up the unploughed rows, simultaneously lifting the ploughed rows.

On overwatered fields use potato diggers KTN-1A, KR-1, KTN-2B, KST-1,4, L-652 with subsequent manual harvesting.

Losses after the combine should be no more than 3% (excluding tubers less than 25 mm), cleanliness - no less than 80, damaged tubers - no more than 12%.

After harvesting, potatoes are sorted on KSP-50 and stored. Seed

potatoes, if impurities are not more than 25%, it is better to store without sorting, so that there is less damage to tubers, then they will be better stored. Potatoes are stored in special storages with active ventilation, with refrigeration units or in bunches with extraction, in trenches. Storage mode: 1) treatment period 3-4 weeks at a temperature of +15 ... +160C, relative humidity 90-95% to heal mechanical damage, remove excess moisture. On tubers thickens the rind, respiration decreases; 2) cooling period 20-40 days with a decrease in temperature per day by 0.50C to +2...+40C; 3) the main storage period at a temperature of
+1.5... .+20C for early varieties and +2.+30C for late maturing varieties and relative humidity of 85-90%. In the last decade, the temperature is reduced to +1.5.+2^{0}C. 1014 days before planting or sale, potatoes are heated by ventilation with warm air +15.+20^{0}C to enhance growth processes in seed potatoes and eliminate sweet taste in market potatoes.

11.5. Promising cultivation technologies

The main methods of potato cultivation according to the traditional Zavorovo technology (after the name of OPF "Zavorovo" of Ramenskiy district, Moscow region) have been outlined above. It provides for several treatments during the growing season, which leads to undesirable consequences: soil compaction by tractor wheels, damage to plants by cultivator, destruction of ridges, tubers coming to the surface and their greening, infection of healthy plants from diseased plants by phytophthora, extraction of weed seeds to the soil surface and their germination in the future. The so-called Dutch technology is more progressive and avoids these disadvantages. The basic elements are as follows: 1) soil milling during pre-planting soil preparation. Milling cultivator with ridge-former "Rumpstad", "Amak", domestic analogues KVK-4, KFG-28 with active working bodies, working from PTO tractor, more thoroughly crush the soil and mix it with fertilizers than KRN-4,2 with passive working bodies; 2) potatoes are planted to a depth of 4-6 cm. On the 15th day after planting, when weeds begin to sprout, form a ridge up to 25 cm high; 3) apply herbicide Zenkor 0,5-1 kg / ha and after that in the field do not drive until harvesting. Soil is not compacted, ridges are not destroyed, tuber nests are not damaged, herbicide film protects from weeds; 4) organic and mineral fertilisers are obligatory; 5) phytosanitary control and spraying with fungicides and insecticides.

11.6. Methods of improving planting material and peculiarities of seed potato cultivation

Potato yields depend largely on the quality of the planting material. As a result of continuous vegetative propagation, potato degeneration can be observed, which is manifested in progressive yield reduction, premature awakening of buds in the eyes, formation of elongated thread-like sprouts, small (often diseased) tubers affected by viral and other diseases. The ecological theory explaining degeneration relates it to unfavourable external growing conditions (high temperature, lack of moisture) during the period of tuber formation. The viral theory attributes degeneration to the viral diseases mentioned above. Physiological aging associated with constant vegetative reproduction is also the cause of this phenomenon.

When growing seed potatoes, by applying certain techniques, it is possible to obtain healthy planting material with higher yielding qualities. Potato seed production is carried out in specialised seed farms on a virus-free basis. At the West Siberian Vegetable and Potato Breeding Experimental Station, elite material is obtained using heat and chemotherapy, as well as the method of in vitro cloning of apex meristem. A 0.1 mm section of apical meristem at the seedling tip from undifferentiated cells, constantly growing and infection-free, is cultured on artificial nutrient medium with the addition of cytokinin, auxin, and hormones for tissue differentiation. As a result, a callus is formed, on which a bud is formed, and from it grows a physiologically young plant free of infection. The material thus recovered is used in seed production to obtain elite tubers.

The following are generally accepted methods for improving the yield quality of potatoes.

1. Seed potato culture on drained peatlands, floodplain soils, on irrigation. These soils are more moist, friable, fertile, characterised by low temperatures of +18...+190C without sharp changes. Tubers from such soils are physiologically younger, form more eyes, have a longer dormancy period, are better stored, have a coarser cork layer.
2. Use of summer planting (late June) of seed potatoes, especially in the south. In this case, tuber formation takes place at a cooler time when there is more moisture.
3. Planting material must be of at least V reproduction.
4. Planting rate for seed purposes is increased (up to 170 thousand/ha), planting depth is increased to 15-16 cm. In this case, a multituber nest of relatively small and physiologically younger tubers is formed, the yield of seed fraction and healthier and more productive plants increases, as sick tubers from a greater depth do not sprout. Densified plantings are less affected by viral infections, as the plants develop faster and acquire age-related disease resistance more quickly.
5. Carry out varietal and phytoclearance treatments during the flowering period.
6. Earlier harvesting dates are practised for seed potatoes, as immature potatoes are less susceptible to infection.
7. Mowing of haulm on seed potatoes is carried out 12-15 days before

harvesting. The outflow of nitrogenous substances is prevented, which protects against physiological degeneration. A thicker peel is formed on the tuber, potatoes are less damaged during harvesting and are better stored.

8. After harvesting, seed potatoes are kept for a fortnight in temporary piles to identify diseased tubers before storage. The tubers are also kept in the light and greened. They release solanine, which is an antiseptic, and the potatoes store better.

CHAPTER 12

12. OILSEEDS

12.1. Use of oils, their most important characteristics

Plant fats are widespread in nature and play a major role in plant life. Oilseeds are particularly rich in fats. The global area planted with oilseeds is about 100 million hectares. The most common oilseeds include soya (52 million ha), peanut (7.5 million ha), rapeseed (15 million ha), sunflower (15 million ha), flax (7.5 million ha), sesame (5.2 million ha). The main areas under crops are in the USA, Canada, India, Brazil, Pakistan, China, Russia.

Compared to proteins and carbohydrates, fats are the most oxidised compounds and therefore have the highest energy content. When burning 1 g of fat, 9500 calories are released, and when burning 1 g of protein - 5500, 1 g of carbohydrates - 4000 calories. Vegetable fats have a number of advantages over animal fats. They are more beneficial to human health, as they do not contain cholesterol. The production of vegetable oils is cheaper. The production of 1 tonne of vegetable oil requires 1 hectare of arable land, while animal oil requires 3 to 10 hectares.

Fats are esters of the triatomic alcohol glycerol with various fatty acids. Fatty acids are divided into two main groups: 1) saturated, or marginal (palmitic, stearic), which at room temperature are in the solid phase and are found in larger quantities in animal fats; 2) unsaturated (oleic, linolenic, erucic), which at room temperature are in the liquid phase and are found in larger quantities in vegetable oils. When unsaturated fatty acids are oxidised, the oil turns into a solid film. This property is used in industry to make olive oil, paints. This requires oils with a large amount of unsaturated fatty acids. An indicator of their content is the iodine number, which is the number of grams of iodine attached to 100 g of fat. It characterises the oil's ability to dry. The higher it is,

the faster the oil dries. According to the degree of drying, oils are divided into the following: 1) quickly drying with iodine number more than 130 (linseed oil, ginger oil, perilla oil), used in the paint and varnish industry; 2) semi-drying with iodine number from 85 to 130 (sunflower oil, rapeseed oil, mustard oil, safflower oil), used in nutrition; 3) non-drying with iodine number less than 85 (castor oil, castor oil, castor oil), used in medicine.

Food and industrial oils should have the smallest amount of free fatty acids (not combined with glycerol), as they complicate the production of oils by necessitating additional processing. An indicator of the free fatty acid content is the acid number, which is the number of milligrams of caustic potassium required to neutralise the free fatty acids in 1 g of fat.

It ranges from 0.1 to 5.7. For grade 1 sunflower seeds it should be no more than 1.3.

Fats serve as raw materials for the soap industry. Fat molecules are broken down by caustic alkali, fatty acid salts are formed, glycerol and water are released. The saponification capacity of fat is characterised by the saponification number, which is the number of milligrams of caustic potassium required to neutralise both free and glycerol-bound fatty acids in 1 g of fat.

12.2. Sunflower

12.2.1. Importance, distribution, yield

Sunflower is the main oilseed crop in our country. The seeds contain up to 50% of semi-drying edible oil with high flavour. It is used for food purposes, for preparation of margarine, canned food, bread and confectionery products. The main fatty acids in sunflower oil are linolenic (55%) and oleic (30%). At present, varieties with a high content of oleic acid have been created, the oil of which has a taste and quality of

is closer to olive oil. It is more resistant to oxidation during storage and heating. Sunflower oil contains vitamins A, D, E, K, phosphatides and other valuable substances. Lower grades of sunflower oil are used to produce soap, paintwork, and oilcloth. By-products - cake, which remains after processing of seeds for oil by press method, and meal - after chemical extraction - are valuable high-protein feed, containing in 1 kg not less than 1 k.u. and 35-40% of protein. Dry baskets with 55-60% yield are also good fodder. The green mass in the flowering phase is used for silage. Sunflower is a good honey crop, yielding up to 30 kg of honey per hectare.

Sunflower is native to the south of North America. It was brought to Europe by H. Columbus in 1510, but at first it was used as an ornamental plant. It was used as an oilseed crop after a serf peasant in Russia, D.S. Bokarev, isolated sunflower oil under a press in 1835.

The area under sunflower cultivation in the world is about 18 million ha (USA, Argentina, European countries). In Russia - about 3 million ha mainly in the North Caucasus, the Central Caucasus, Volga region, Tatarstan, Chuvashia, Western Siberia. The average yield in the country remains low, about 0.8 tonnes/ha. On varietal test plots, in advanced farms on high agrophonous soil yields of 2.5-3 tonnes/ha are obtained.

12.2.2. Classification and botanical characterisation sunflower

The annual sunflower (Helianthus annuus L.) belongs to the Asteraceae family. There are two species of sunflower: cultivated sunflower (Helianthus cultus Wenzl.) and wild sunflower (Helianthus ruderalis Wenzl.) The cultivated sunflower is divided into two subspecies: cultivated sowing sunflower (ssp. sativus Wenzl.) and cultivated ornamental sunflower (ssp. ornamentalis Wenzl.).

Sunflower varieties are divided into groups depending on seed size, oil content and huskiness:

1) Oilseeds - small seeds 8-14 mm long, weight of 1000 seeds 35-75 g, with low huskiness (22-36%), large kernel, which almost completely fills the cavity of the seed, with a fat content of 53%;

2) gryzovy - large seeds 15-25 mm long, 1000 seeds weigh 100-170 g, with high huskiness (42-56%), the kernel fills the seed cavity poorly, oil content is low (20-35%). Gryz varieties have taller, larger plants, so they are recommended to be grown for silage;

3) mezheumecs occupy an intermediate position (Fig. 14).

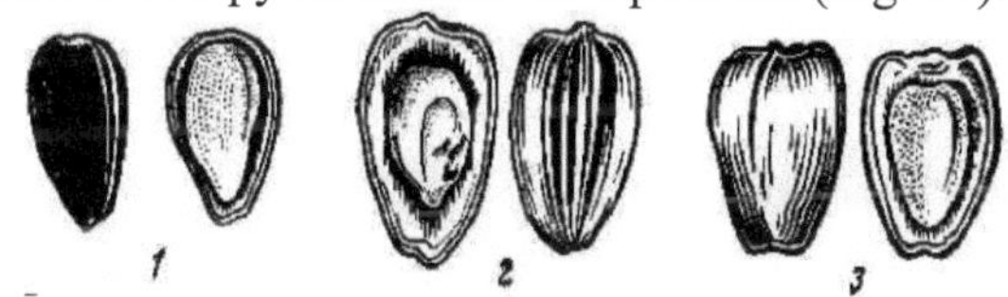

Figure 14: Sunflower seeds:
1 - oilseed; 2 - rodent; 3 - mezheumk

Sunflower is an annual plant. The root system is rod-shaped, powerful, penetrating to a depth of 2-4 m, 100-120 cm to the sides. The stem is erect, woody, made by a loose core, 0.7-2.5 m high, non-branched. Leaves are simple on long petioles, large, heart-shaped-oval with serrated edges, densely pubescent. One plant has 15 to 30 leaves, depending on the variety. Inflorescence is a basket with a diameter of 10-20 cm in oilseeds, 40 cm and more in grasses. The basket is surrounded by a leaf wrapper and consists of a peduncle with orange-yellow tongue-shaped flowers, which are sterile and serve to attract insect pollinators (Fig. 15). Almost the entire peduncle is occupied by tubular flowers, which can number from 600 to 1200 in one basket. Tubular flowers have a pistil with a lower ovary and a column, the corolla is yellow or orange. The stamens are five with free filaments and fused anthers.

The tubular flowers set fruit. Sunflower is a cross-pollinating, entomophilous plant, pollinated by insects. The fruit is an ovoid-shaped seed, consisting of a seed (kernel) with a thin seed coat and a leathery pericarp (rind). In the pericarp under the epidermis, between the cork tissue and sclerenchyma in shell varieties, there is a shell layer of cells, in which a black carbon, insoluble in water, acids and alkalis, substance (phytome-lan) is formed, consisting of 76% carbon (Fig. 16). This layer protects sunflower seeds from penetration by sunflower moth larvae. All modern varieties are shell-shaped.

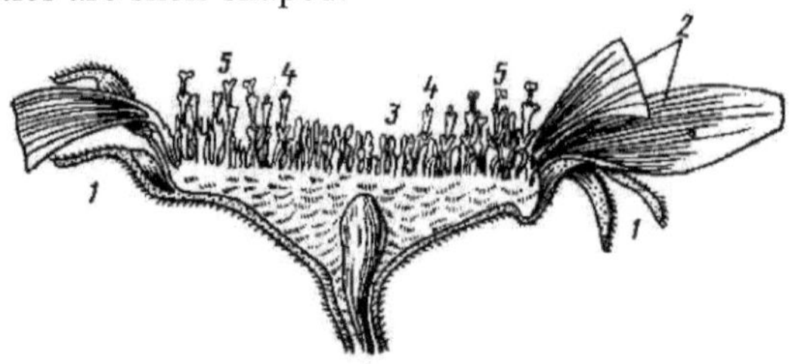

Figure 15. Structure of the sunflower basket:
1 - wrapper leaves; 2 - reticulate flowers;
3 - unopened tubular flowers; 4 and 5 - tubular flowers

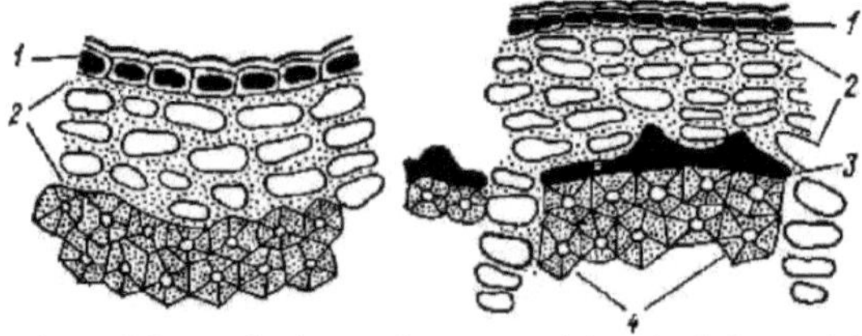

Figure 16. Section of the peel of a sunflower seed (on the left, panicle-less, on the right, shell-less):
1 - epidermal cells; 2 - cork tissue; 3 - shell layer;
4 - sclerenchyma cells

The husk is the fruit shell or rind. Seed huskiness is the weight of husk expressed as a percentage of the seed weight. The most valuable for oil production are
oilseed varieties with low huskiness (about 20%). The weight of 1000 seeds is from 40 to 125 g. The seed consists of an embryo and a seed coat. The embryo consists of a germinal root, a stalk and two seed pods, which emerge on the soil surface during germination.

12.2.3. Biological characterisation of sunflowers

Sunflower is native to the south-western part of North America, the dry, sultry prairies. The process of domestication also took place in steppe conditions. Sunflower is therefore a drought-tolerant plant that can tolerate drought and high temperatures well.

Minimum seed germination temperature is +4...+60C, optimum temperature is +15...+200C. Sprouts withstand short-term frosts up to -8^{0}C. The most favourable temperature for growth is +25...+270C. Temperatures above +300C depresses plants. In the flowering phase -1...-20C causes death of flowers.

Sunflower is a drought-tolerant plant due to its strong root system, which utilises moisture from layers inaccessible to other crops. The stem and leaves are pubescent, which protects against excessive overheating and evaporation from the leaf surface. It is a typical mesophyte, evaporating quite a lot of moisture. The transpiration coefficient is between 400 and 570. Sunflower consumes moisture unevenly: sprouting - basket formation - 23%, basket formation - flowering - 60, flowering - ripening - 17%. In the initial period before basket formation, sunflower tolerates drought, both soil and atmospheric, well. The critical period in terms of moisture is basket formation - flowering - filling. Lack of moisture at this time leads to hollowness in the centre of the basket, as well as underdevelopment of seeds and their stubbiness.

Sunflower is a short-day plant, so as the crop moves northwards, the growing season increases. It is a very light-loving plant. It is characterised by heliotropism, i.e. the ability of the basket to rotate behind the sun, so it always faces the sun. Towards the end of flowering, the rotation stops and the basket always faces east. If the rows are orientated from south to north, the basket will always face the inter-row. In this case, the plants shade each other less, are better ventilated, have less disease, and there are fewer losses when harvesting because the basket is hit from the side.

The best soils for sunflower are chernozems of medium granulometric composition, as well as chestnut soils. Favourable pH = 6.8. Heavy clay soils as well as light sandy, waterlogged, strongly acidic or saline soils are unsuitable. The root system of sunflower is characterised by an increased digestive capacity, so it is not very demanding on soil fertility.

In the process of sunflower growth, the following phases are noted: seed germination, emergence of seedlings (noted with the appearance of seed pods on the soil surface), first and second pair of leaves, third and fourth pair of leaves (at this time the basket begins to form), budding (the beginning is noted with the appearance of a basket with a diameter of 2 cm), flowering, seed growth, seed filling (36-40% moisture content), ripening (physiological ripeness, back of the basket is yellow, seed moisture 18-20%), full ripening (economic ripeness, basket is brown, seed moisture 12-14%).

According to the length of the vegetation period sunflower varieties are divided into the following: 1) medium-ripening (120-140 days); 2) early-ripening (100-120 days); 3) early-ripening (80-100 days). In Altai Krai, only early-ripening, early-ripening and mid-early varieties are zoned.

12.2. 4. Sunflower cultivation technology

Place in crop rotation. The best precursors for sunflower are winter and spring grain crops, maize, clean and occupied fallow. Sunflower can be returned to its former place not earlier than in 7-8 years to prevent the development of diseases and pests. In steppe areas of Altai Krai, where crop rotations with short rotation are practiced, the most rational is the following placement of sunflower: 1) fallow; 2) spring wheat; 3) spring wheat;

4) sunflower; 5) oats. In this crop rotation sunflower occupies % of the field, which makes it possible to return sunflower to the same place after 8 years in a short rotation. This is achieved by changing sunflower and oat fields in rotation. Sunflower should not be placed after sugar beet, alfalfa, Sudan grass, which dry up the soil, as well as after rape, soybeans, peas, beans, as these crops have common diseases with sunflower (false powdery mildew, grey rot).

Tillage. The main tillage is carried out by ploughs KPG-2-150, OPT-3-5, KPSH-9, combined aggregates (SMARAGD, APK-7,2) to a depth of 20-22 cm in steppe and up to 25 cm in forest-steppe. In spring, when the soil is physically ripe, harrowing and levelling are done with toothed or needle harrows, harrowers with rollers, and after ploughing - with draggers. Levelling allows to apply herbicides more evenly, to make more technologically advanced sowing, levelled by depth. Before sowing, pre-sowing cultivation is done to a depth of 6-8 cm with simultaneous rolling.

Application of fertilisers. When forming 1 kg of seeds sunflower takes out 5-6 kg of nitrogen, 2 kg of phosphorus, 10 kg of potassium. Additional nitrogen in combination with other elements enhances growth, increases leaf surface, usually slightly reduces oil content. Phosphorus increases the number of reproductive organs, accelerates development, increases drought resistance, increases oil content. Together nitrogen and phosphorus are more effective than separately. Potassium both unilaterally and in combination with nitrogen and phosphorus does not give a significant increase in yield on chernozems and other soils where there is enough potassium. The effect of additional potassium is observed only on soils where it is not enough - grey forest soils, podzol soils, meadow-chernozem soils.

Sunflower does not respond much to high fertiliser rates due to weak activity of enzymes regulating nitrogen metabolism. Medium rates of $N_{40}P_{60}$ are effective on chernozems, and $N_{40}P_{60}K_{40-60}$ on potassium-poor soils. We can expect a yield increase of up to 0.2 tonnes/ha. Increasing the rate above the recommended does not increase yield, but reduces oil content by 2-3%.

If you use a more effective method of fertiliser application - local-tape, it is possible to reduce the application rate by 2 times to ¹°Ðz° d.v./ha, and yield increase is obtained up to 0.3 t/ha.

Sowing. Calibrated seeds are used for sowing, which allows to obtain more aligned plants and reduce losses at harvesting. The use of heavier seeds (with a weight of 1000 seeds of at least 80 g for varieties and 50 g for hybrids) significantly increases yields. To prevent diseases (white and grey rot, etc.), seeds are treated no later than two weeks before sowing, using, for example, Apron Gold KE (2 kg/t seeds mixed with microelements).

Seeds of high oilseed crops, including sunflower, are characterised by higher requirements for heat during germination. Sunflower sowing starts at the temperature in soil at the sowing depth of +8...+100C. In Altai Krai it is the end of the first - beginning of the second decade of May. Earlier sowing is preferable in arid areas. It is important to link the sowing dates with the possibility of weed eradication. On clean fields, as well as when soil herbicides are applied, it is necessary to sow as early as possible, on weedy fields - later.

Sunflower is sown in a wide-row dotted method with row spacing of 70 cm (less often 45 cm) using pneumatic seeders SUPN-8, SPCH-6, SKPP-12, STV-8 with harrows and harnesses, OPTIMA, MONOSEM. Sowing depth - 6-8 cm, in dry conditions - 8-10 cm. The optimum plant stand density for harvesting in the Kulunda steppe is 4° thousand/ha, in the forest-steppe - 5° thousand/ha. When calculating the seeding rate, it should be taken into account that the field germination of seeds is 15-25% less than laboratory germination. In addition, it is necessary to take into account the fallout of plants during their care. For each tillage after sprouting it is necessary to increase the seeding rate by 5 per cent. As a result, the weight seeding rate can be from 5 to 8 kg/ha. When growing sunflower for silage, the plant density should be 200-250 thousand/ha.

Sowing care. When sowing in loose soil it is necessary to roll immediately after sowing with ring-spur rollers. In intensive technology herbicides are used to control weeds. Soil herbicides are used if the soil is well moistened, they act at germination of weed seeds, so they are introduced into the soil before or after sowing with embedding into the soil. During sunflower vegetation often apply anti-grass herbicides Fusilad forte, Furore super. The use of herbicides allows to refuse inter-row cultivation on light soils or reduce their number on heavy soils. In herbicide-free technology, harrowing and inter-row treatments are carried out. Harrowing before sprouting is effective 4-5 days after sowing in the phase of white threads of weeds tooth harrows BZSS-1,0, BP-0,7, and in fields with large amounts of plant residues - rotary hoe MRN-8,4 across the rows. Harrowing on sprouts destroys annual late weeds. It is carried out in the phase of 2-3 pairs of sunflower leaves in the afternoon across the rows. Fallout should be no more than 5%. Inter-row cultivation is carried out by KRN-4,2, KRG-5,6 with lancet and single-sided tines, with KLT-360 harrows to cover weeds in the row zone, weeding harrows KLT-38 to loosen the soil in the row zone before sprouting. Start the first treatment when sunflower plants reach a height of 20 cm, to a depth of 68 cm, the second treatment - at 8-10 cm after 10-15 days, the third - at 6-8 cm when weeds appear. Treatments are stopped when plants reach 60 cm.

In the flowering phase, additional bee pollination is carried out, 3 bee-families per 1 hectare. This increases the yield by 0.2-0.3 tonnes/ha.

Sunflower harvesting begins at the phase of economic ripeness when seed moisture is 12%, when most baskets are brown in colour. When sunflowers are over-mature, when the moisture content drops to 8%, losses from shattering at the root increase, especially in steppe areas. Hybrids ripen more quickly than varieties, so harvesting them 5-7 days earlier to prevent shattering. In the forest-steppe zone in Siberia, in conditions of cold rainy autumn, harvesting often has to start at high humidity (18-20%). In this case, pre-harvest desiccation with magnesium chlorate 20 kg/ha with a liquid consumption of 100 litres/ha by aerial method is effective 40 days after flowering, when the seeds are full. In this case, in 10 days the moisture content is reduced to 12%, maturation is accelerated, harvesting time is shortened, and losses from diseases are reduced. The crops should be harvested in 5-7 days.

Harvesting is carried out by combine harvesters equipped with special devices PSP-1.5, PSP-8, PSP-10, the set of which includes a special windless harvester, which cuts only baskets without stalks on a high cut, as well as a basket and stalk chopper. Row dividers feed the stalks to the rotor blade of the cutterbar without impact. A vibrating conveyor ensures a steady flow of baskets and shattered seeds into the auger chute. To prevent seed collapse, the drum speed is reduced to 425-450 revolutions per minute and to 300 revolutions per minute in the seed sections.

Seeds coming to the current contain a lot of impurities and have high humidity. Moist seeds left even for a day are self-heating, which leads to their spoilage. Oil from such seeds has an increased acid number. Seeds are cleaned on OVP-20, OS-4,5A, ZAV-20, etc., dried to a moisture content of 7-9%. Heating temperature of seeds should be no more than 40-45^{0}C. Seed filling for long-term storage without active ventilation is allowed at humidity not more than 7%.

12.2.5 Botanical and biological characteristics of oilseed plants cabbage family

In addition to sunflower, cold-resistant oilseed plants of the cabbage family (Brassicaceae) can be of great importance in Siberia: 1) rapeseed

(Brassica napus L. ssp. olifera Metzg.), which is represented in culture by two forms: winter - biennis and spring rape, or kolza - annua; 2) turpentine also has a winter form (Brassica rapa L. ssp. olifera D.C.) and spring (Brassica campestris L.). In the world practice, rape and turpentine are often combined under the common name "rape" or in southern Europe "kolza"; 3) blue mustard (Brassica juncea Czern.); 4) white mustard (Sinapis alba); 5) spring ginger (Camelina sativa); 6) crambe (Crambe abyssinica) (Tables 34, 35).

Characteristics of oilseeds of the cabbage family

View	Shape and surface	Colouring	Flavour	Oslimation in water	Weight of 1000 seeds, g	Oil content,%	Iodine number	Utilisation
Bluestem	Oval-rounded, clear-mesh.	Dark brown	Stinging with an essential odour	They don't slime	1,7-4	35-49	120-121	Food, mustard powder
White mustard	Globular, thinly netted.	Cream	Bitter with no essential odour ha	They're very slimy	4-6,5	30-40	92-122	Food
Rapeseed	Globular, slightly cellular.	Black	Grassy	They don't slime	3-6,9	33-50	101-107	Technical, food
Carrot-top	Oval with longitudinal groove	Orange-yellow	Turnip flavour	They're waxing	0,9-2,5	32-46	139-157	Olefovare
Surepea	Almost globular, coarsely netted.	Reddish-brown.	Grassy	They don't slime	2-3	34-38	100-103	Technical
Crumbe	Almost spherical	Straw yellow	Grassy	Same	5-6,5	34-53	93-97	In the food industry

Main features of oilseed crops of the cabbage family

Signs	Bluestem	White mustard	Rapeseed	Carrot-top	Surepea	Crumbe
Stem	Bare or downy bluish in the lower part	Draped with bristly hairs, purple at the base	Naked, blue	Loosely pubescent with short hairs and long bristles, green or faintly blue	Naked or pubescent underneath, light bluish	Bare or pubescent in the lower part, bluish
Shape of root leaves	Lyre-shaped-pinnate-sectioned, less often entire	Lyre-shaped-peri-cut	lyre-shaped-pe- riston-cut.	Lancet	lyre-shaped-pistonad-incised.	Rounded
Upper lobe of the leaf	Large, semi-oval	Large, wide-oval	Large, oval, blunt.	-	Large, oval	Large, ovoid
Tumescence, colouring	Pubescent or less often glabrous, light bluish	Stiff-haired, green	Covered with a waxy plaque, bluish.	Lightly pubescent, green	Downy, green	Unfurred, light blue colour.

Inflorescence and flowers	Shield-shaped, bright yellow.	Cyst-shaped yellow	Cyst-shaped, light yellow.	Cystic, pale yellow.	thyroid, golden yellow.	Loose brush, white
The fruit is a pod	Linear, 4-faceted, slim, with short spout !4 flap lengths	Straight or curved, articulate with a long sword-shaped spout equal to the length of the flaps	Narrow, straight or bent, with a thin short spout, pods at right angles to the stem	Obovate, with a short spike-shaped spout	Narrow, straight or bent, with an elongated conical spout	Globular without spout

Seeds of these crops contain oil from 30 to 50%. Seed yields are from 0.8 to 1.2 tonnes/ha. The utilisation of the oils is shown in Table 33. Rapeseed oil was not used for food until recently because of its high content of erucic and eicogen fatty acids and glucosinolates. Nowadays, erucic acid-free varieties have been developed with erucic acid content of 0-5% against 37-50% in old varieties.

This has significantly increased the demand for rapeseed oil for food purposes as well as for technical purposes as a source of biodiesel, a renewable, environmentally friendly fuel.

By-products after oil extraction from seeds - cake, meal - contain up to 40% protein and are used for livestock feed, but it is better to use it in the form of additives, as glucosides in large quantities can cause inflammation of the intestines and kidneys. Green mass in the flowering phase of rapeseed, white mustard, turpentine is used for livestock feed. In 100 kg of green mass contains 15 k.u.d. Yield of green mass is up to 20 tonnes/ha, and in winter forms - up to 30 tonnes/ha.

Biological characteristic. All these plants are cold-resistant, moderately demanding to heat. The minimum temperature of seed germination is +2...+30C, the optimum temperature is $+15^{(0)C}$. Sprouts can withstand frosts up to -5^0C, and in autumn the green mass can withstand frosts up to -8^0C. These crops have a short vegetation period: mustard blue - 90-100 days, mustard white - 65-90, rapeseed - 95-100, ginger, turpentine - 75-85 days.

The listed crops are long-day, moisture-loving plants. In dry years rape, white mustard significantly reduces the yield. Blue mustard, redcurrants, turpentine are relatively more drought tolerant compared to white mustard.

The highest yields are obtained when sowing rape and mustard on chernozems. White mustard, surypitsa and red mustard are less demanding to soil fertility. The root system of white mustard is characterised by a high digestive capacity, so it can be grown on poor and slightly acidic soils. The cleansing role of plants of the cabbage family, especially white mustard, in relation to pathogenic microflora of the soil is known. After rape, wheat is less affected by root rots. Having a short vegetation period, they quickly form a high yield of green mass, well suppress weeds, so they are often recommended as intermediate, sideral crops.

Cultivation technology. For the cultivation of oilseed crops of the cabbage family, good predecessors are winter and spring cereals, row crops, legumes. These crops respond well to the introduction of organic and mineral fertilisers. Recommended rates - N45-60P45-60K45-60.

Tillage for these crops is similar to that used for early spring crops. Since the seeds are small, the pre-sowing treatment is more demanding. There should be no lumps, furrows or ridges on the surface. The soil should be levelled and moist. For processing use aggregates AKP-2,5, USMK-5,4, depth - 4-5 cm. Treated seeds are used for sowing. Sowing dates for seeds are early, in early May. Early crops better use the moisture reserves in the soil, less affected by cruciferous fleas, as they have time to get stronger before their mass appearance. Plants from early sowing dates have a longer period of laying generative organs, so they are more productive. In addition, they ripen more quickly and have a higher oil content. For fodder purposes in the green conveyor belt can be sown from early spring until mid-July.

The sowing method is row sowing. Seeders: grain-grass (SZT-3,6), AMAZONE, as they sow small-seeded crops more qualitatively. When sowing with grain seeders it is possible to use ballast - superphosphate 20 kg/ha, mixing it with seeds just before sowing.

Seeding rate - 2.5-3 million/ha of germinated seeds (10-12 kg/ha), which provides about 200 plants per 1 m^2 by harvesting. Sowing depth is 2-3 cm.

Plant care. Rolling of dry loose soil before and after sowing, application of herbicide (Treflan - 2.5 l/ha) under pre-sowing cultivation, harrowing before and after sprouting (in the phase of 3-4 leaves), control of diseases and pests. Systemic protect seeds and seedlings not only from diseases, but also from pests at the beginning of development systemic dressing agents Maxim and Kruiser protect well.

Harvesting for seed is complicated by non-simultaneous maturation of pods. Separate harvesting reduces losses from shattering, moisture and contamination of seeds, and improves sowing and technological qualities of seeds. Swathing is started when the leaves fall off and the pods on the main branch turn yellow and the seeds in them acquire a characteristic colour. Plants are mown with ZhVN-6A, ZhRB-4,2A cutterbars at a height of 15-20 cm. Windrows are threshed at a seed moisture content of 10-12% at a threshing drum frequency of 700-800 rpm. The combine is equipped with a device PKK-5 for threshing small-seeded and cereal crops, which allows to reduce losses and crushing of seeds, as well as PPT-3 (web-transporter picker), which helps to reduce losses from shattering. For rape harvesting there is a special attachment to combines - PZR-6.

White mustard pods do not crack, so it can be harvested directly. Cleaned seeds are stored at a moisture content of no more than 10%, or at a moisture content of 7% for long-term storage. When harvesting for green mass, the plants are cut at flowering stage.

CHAPTER 13

13. LEN

13.1 Utilisation, distribution, yields

Flax is one of the best spinning crops. It is grown for its natural fibre, which is contained in the flax stalks. The fibre content is 18-33% in linseed flax, and 2 times less in mulberry flax. Flax seeds contain up to 35-39% oil (up to 42-44% in oilseed flax) and up to 25% protein.

Linen fibre has high technological properties and is the main raw material resource for the textile industry in Russia, as cotton fibre from Central Asia has become much more expensive. Linen fibre is 2 times stronger than cotton fibre, 3 times stronger than woolen fibre and slightly inferior to silk yarn. Products made of linen are good, beautiful, hygienic, do not cause allergies. From 1 kg of linen fibre 10 m^2 of batiste, 2.4 $m^{(2)}$of cloth or 1.6 m^2 of tarpaulin, as well as twine, threads, ropes, and from indirect fibre - hemp. Only long flax fibre No. 12-14, which can be obtained from linen straw No. 1.5-2 and higher, is used for making cloth. The quality of the fibre reflects the number, which means the number of skeins of yarn of a certain length (245.5 m) from a unit mass of fibre (453.6 g). The higher the number, the higher the quality, the lower the fibre consumption to produce 1 $m^{(2)\ of}$ fabric. For example, for batiste, the fibre number is 150.

Bark, which contains 60% cellulose, is used to make cardboard, building boards, ethyl alcohol, acetone. Oilseed flax yields indirect fibre, which is used for making twine, rope and twine. The oil is used for making olive oil, paints, varnish. Cake and meal contain 30% protein, are valuable fodder. In 1 kg of cake - 1.15 k.u.

The flax straw harvest contains 75% flax straw, seeds - 10-15%, chaff - 10-15%. Thistle yield from flax straw harvest is 70%.

Flax has been cultivated in Russia since the 3rd to 5th centuries AD. Flax has been the main export product in Russia since the times of Peter the Great. Pskov, Novgorod and Kashin flaxes are known.

60% of the world's flax sown area is in Russia, Belarus and Ukraine. The area sown in the world is 1 million hectares, in Russia - 200 thousand hectares, mainly in areas with temperate climate in the Non-Black Earth Region, while oilseed flax crops are widespread in the Central-Central Zone, Western Siberia, and the Volga Region.

The average yield of flax fibre in the country is about 0.4 t/ha, flax straw - 2.0 t/ha, while the potential yield reaches up to 1.6 t/ha of flax straw.

In Altai Krai, the area under flax planting is 4 thousand hectares, mainly in the foothill regions with moderate and more humid climate (Biysky, Smolensky, Zalesovsky, Zarinsky, Soltonsky, Togulsky and Tselinny districts). The flax straw yield is 2.3 t/ha, fibre yield is 0.47 t/ha, which is higher than the national average. Flax-cuddyash occupies an area of 4.5 thousand hectares, its yield is 0.14 tonnes/ha of seeds.

13.2 Morphological characteristics of flax plants and classification

Flax is an annual herbaceous plant. Seeds are flat, ovoid, smooth, shiny, brown. The weight of 1000 seeds is 2.8-6 g. Sprouts bear seedpods on the surface. The true leaves are lanceolate, sessile, arranged in order, dying off during seed ripening. Stem up to 1 m or more tall, cylindrical, thin, covered with waxy plaque. Inflorescence is an umbrella-shaped brush. The flower is of the five-celled type (5 sepals, 5 petals of blue, pink, white colour, 5 stamens, five-celled ovary with five columns). Plants with blue flowers are more productive. Flax is self-pollinating. It blooms for 6-10 days. The fruit is a spherical boll, which opens when it stops blooming. The root system is tap root, underdeveloped, 8-10% of the plant weight. 80% of roots are located in the arable soil horizon.

The stems, in particular the bast, the peripheral part of the stems, are harvestable. The fibrous bundles of the bast consist of elementary fibres glued together by pectin. Each fibre is a strongly elongated cell of conducting tissue with length from 20-30 mm to 100 mm and 80 μm in diameter. There are 10-50 fibres in a bundle. The best quality fibre is in the middle part of the stem. In thin and long stems (diam.
no more than 0.8-1.2 mm and a height of at least 70-80 cm) a high quality fibre is produced, which should be long and thin.

Cultivated flax (Linum usitatissimum L.) belongs to the flax family (Linaceae). Linum crepi- tans (Linum crepi- tans) occurs in cultivated flax crops as a weedkiller, characterised by the fact that mature bolls open wide and seeds with poorly developed spouts fall off.

Common cultivated flax is divided into subspecies: Eurasian, Mediterranean, and intermediate. The latter two are distinguished by larger bolls and seeds. The Eurasian subspecies, which is divided into groups of varieties, is cultivated here (Table 36).

Table 36

Flax variety groups

sign	Long flax V. elongata	Flax mezheumok V. intermedia	Curly flax V. brevimulti-caulia	V. prostate
Height, cm	70-120	50-70	30-50	50-60
Branching	It doesn't branch	Weakly branched	Branching out from the base	Strongly branched
Number of stems	1	1-2	4-5	4-6
Number of boxes	8-10	15-25	30-50	30-40
Weight of 1000 seeds, g	3-5,5	4,5-6	5,0-8,0	2,5-5
Oil content, %	35-39	38-42	38-45	40-42

Biological form	Yarovoy	Yarovoy	Yarovoy	Winter

13.3 Peculiarities of flax biology

Flax is a long-day plant originating from temperate latitudes, which explains much of its biology. Flax is favoured by moderate spring and summer temperatures with intermittent rain and clear weather. The minimum seed germination temperature is +2... .+5^{0}C, the optimum temperature is +150C. Sprouts withstand frosts up to -3... .50C. The sum of active temperatures during the growing season is 1100-1500^{0}C. Optimum temperature for growth is +16.+18^{0}C.

Temperature over +22^{0}C depresses growth in height, increases branching, so the quality of fibre deteriorates. The vegetation period is 75-100 days.

Flax cucurbits are more demanding of heat. During flowering and ripening, the optimum temperature is +20.+22^{0}C. The sum of active temperatures is 1600-1800^{0}C.

The following phases are noted in the development of lingon flax: sprouting, herringbone phase, budding, flowering, ripening.

Flax is a moisture-loving crop due to its weak root system. Seed swelling requires a lot of moisture - 160% of the seed weight. In the initial period flax grows slowly, but from the "herringbone" phase, before and during budding, it grows vigorously and its moisture requirements increase. Flax is particularly demanding of moisture during budding and flowering. Thereafter, growth in height stops. Excessive moisture during flowering and ripening causes lodging and disease damage. Flax should not be grown in areas where groundwater is close to the soil. The transpiration coefficient is 400- 500.

Flax is characterised by moderate light requirements. Strong light increases branching and reduces the yield of long fibre. But excessive shading causes lodging.

In the Non-Chernozem zone flax is traditionally grown on cultivated, sod-podzolic soils, medium and light loamy, slightly acidic with pH = 5.6-6, with humus content not less than 2%. Heavy clayey, acidic peaty soils are of little use. In Altai Krai flax is grown on leached, podzolised and ordinary chernozems.

13.4. Cultivation technology

Place in crop rotation. Flax can be returned to its former place after 7-8 years. In case of repeated cultivation, flax fatigue is observed, i.e. a sharp decrease in yield, as pathogenic microorganisms (fusarium, anthroknosis, polysporosis, etc.), toxic substances accumulate, one-sided soil depletion manifests itself, specialised weeds (flax chaff, flax thorn, flax redworm, wilt) accumulate. Perennial grasses are a good precursor for flax, especially when cultivated in traditional flax-growing areas with low soil fertility and where no fertilisers are applied. On clean fields, when fertiliser is applied, the importance of such predecessors as winter and spring cereals, legume-oat mixtures increases, as flax after these predecessors is more level and technological than after grasses. Wheat, potatoes, beetroot are sown after flax.

Fertilisation. Flax is demanding for available mineral substances, as it has an underdeveloped root system with a low digestive capacity, and the bulk of nutrients are supplied to the plants in a short period of time. By the beginning of flowering, 84% of nitrogen, 80% of phosphorus and 70-90% of potassium are consumed. The output per 1 tonne of straw and the corresponding number of seeds is 10-14 kg of nitrogen, 4.5-7.5 kg of phosphorus and 11-17.5 kg of potassium. Nitrogen promotes growth and increases the yield of long fibre. Phosphorus promotes root system development, accelerates maturity, increases seed and fibre yields. Potassium increases lodging resistance and fibre yield.

Organic fertilisers are not applied directly to flax to avoid crop variegation and weediness, as well as the formation of coarse fibre.

On fertile soils, if the grain predecessor yield was at least 2.5 t/ha, 25 kg d.w/ha nitrogen is applied to flax, if less - 30 kg d.w/ha. Complete mineral fertiliser is applied to flax in the ratio NPK 1:2:2.

On soils with pH = 4.5, liming is necessary. But flax is sensitive to liming. On freshly limed soils, flax suffers from bacteriosis and physiological wilting, as boron and potassium become less available, which leads to disturbances in the formation of the conductive system, and the fibre is coarse and brittle. Therefore, the soil should be limed under the predecessor or in a fallow.

Tillage. Flax requires more thorough tillage due to its weak root system and the fact that it is a small-seeded crop. The layer of perennial grasses is discredited in two directions with heavy harrows, then 22-25 cm of mouldboard ploughing is done. After stubble forerunners, husking and ploughing are done. Early spring cultivation is better to be done with disc harrows in order not to turn out the turf. Pre-sowing cultivation is done with machines combining fine loosening, levelling and rolling (VIP-5, RVK-3,6).

Before sowing, seeds are dressed and heated by air-thermal heating. Sowing is started when the soil at a depth of 10 cm warms up to +6...+80C (second decade of May).

To obtain plants with thin and long stems with a high content of high-quality fibre, flax should be grown in a thickened state (1500-1600 plants/m$^{2)}$, therefore the seeding rate is 18-25 million/ha of germinating seeds. In order to evenly distribute the seeds, they are sown in narrow-

row method with row spacing of 7.5 cm using seed drills SZL-3.6A, SLN-48A.
When germinating flax seedlings come to the surface, so it does not tolerate deep sowing. The sowing depth is 1.5-2 cm, on light soils - 3 cm. If flax is to be harvested with straw spreading on the flax bed, red fescue, pasture ryegrass or creeping clover 10 kg/ha should be sown under it. By the time of flax harvest, the grasses form a grass stand up to 20 cm high, ensuring that the straw is isolated from the ground during flax spreading and optimal curing conditions. Grass seeds are mixed with flax seeds before sowing and sown together.
Crop care. After sowing, it is necessary to carry out post-sowing consolidation and soil crust elimination. Flax grows slowly at the beginning of vegetation, so it is important to control weeds, but this crop is sensitive to most herbicides. It is best to use herbicides when the flax plants are 58 cm tall and in the herringbone phase. At this time there is maximum waxy patina on the flax leaves, herbicides run off the plants. Herbicides can be applied in a mixture with ammonium nitrate or urea - 10 kg/ha with the addition of microelements (boron - 0.25 kg/ha, zinc, molybdenum - 0.1 kg/ha each).
Harvesting and primary processing. To obtain high quality fibre, harvesting begins in the early yellow ripeness phase, which occurs 25-30 days after mass flowering. Signs of early yellow ripeness: stems are yellow, leaves are 2/3 shattered, bolls have greenish veins, seeds are in the phase of waxy ripeness. In 5-7 days after early yellow ripeness comes yellow ripeness, the bolls are yellow, the seeds harden and acquire their characteristic colouring. At full ripeness the stems and bolls are brown, the seeds in the boll are ripe and make a noise when shaken. Fibre at yellow and full ripeness is coarser, woody and loses elasticity.
In seed crops, harvesting begins at the yellow ripeness phase.
Flax is harvested in two ways. The first is by spreading straw on the flax bed to produce flax cod. The flax harvester LK-4A with a spreading machine and a combing device performs treading (pulling out with the root) of plants and combing of seed bolls. The pile with bolls goes into the trolley, and the flax straw is spread on the flax bed on the sown grass. The straw should be 8-10 cm above the soil. During curing, the straw should be turned 3-4 and 10-20 days after spreading. OSN-1 wrapper is used. During spreading, flax straw turns into a shale cod as a result of the aerobic fungus Cladospo- rium herbarum Zn. The so-called "dewlap" of the straw occurs. The best conditions for curing occur in August with warm weather (+18^{0}C) and abundant dew for 3-4 weeks (in case of late spreading 5-7 weeks). The stems become grey, pectin

substances are destroyed and it becomes possible to separate the bark (woody part of the stem) from the husk (fibres). To determine the end of curing, samples of tresta - "torture" - are taken in the field. Tresta is passed through a laboratory crusher and rubbed. At under-curing the fibre is difficult to separate from the bark, at over-curing there is a partial separation of elementary fibres from each other. After maturing dry cod with humidity not more than 20% is lifted and knitted into sheaves by picker PTN-1 or formed into bales by baler PRP-1,6 with subsequent loading and sending to the flax mill. At high moisture content of the cod, it is knitted into sheaves and set for drying in cones or tents.

After combing, the raw pile with bolls has a moisture content of 30-65%. It is dried in floor dryers, laid in a layer of 0.7-1 m, to a moisture content of 16-18%, and then threshed on stationary threshers MV-2.5A. Then seeds are cleaned on seed cleaning machines. During long-term storage, seed moisture content should not be more than 10%.

The second method - on weed-free fields, with levelled and not lodged crops, the following technological complex is used for flax harvesting: treading, combing of bolls, bundling of flax straw into sheaves. The combine harvester LKV-4A equipped with a sheaf binder works in this process. Drying of sheaves in "babki" 6-10 days, picking up, loading of sheaves and transporting to the flax factory. In this case, the straw is produced at the flax mills in the process of water lobe in heated soaking pools. Sheaves of straw are loaded into the pools and poured with warm water 36-38^{0}C for 6-9 hours, then kept in the flow of warm water for 3-5 days, squeezed and dried. Pectin substances are decomposed under the action of anaerobic bacteria Bacillus felsi- neus Carbone.

13.5. Peculiarities of oilseed flax cultivation

Mezheumok flax and less frequently curly flax are used for oilseed. They are less demanding of moisture and soil fertility, but more demanding of heat than longan flax. Flax can be grown for oilseed both in the forest-steppe and in the steppe on chernozem and chestnut soils.

Oilseed flax is sown in a row or narrow-row method at a rate of 12-15 million/ha (50-80 kg/ha). For two-sided use for oilseed and baling, the seeding rate is increased up to 90 kg/ha. Plants are torn at the phase of yellow ripeness, knitted into sheaves, dried, threshed on flax threshers. If the stalks are not used, mowing and threshing are carried out at the beginning of full ripeness by combine harvesters on a low cut.

CHAPTER 14

14. PERENNIAL FORAGE GRASSES

14.1. Importance, utilisation, yield perennial forage grasses

Perennial forage grasses play a major role in creating a solid fodder base for livestock farming. Perennial plants, unlike annual plants, annually regrow in spring from the buds laid in the renewal zone at the expense of nutrient reserves formed in the previous year. In this regard, perennial crops have a number of advantages. Having a longer growing season, they provide fodder from early spring to late autumn. These plants make maximum use of the sun's energy and form a large biomass, leaving up to 10 tonnes/ha of organic matter in the soil, including 120-150 kg/ha of nitrogen. Perennial leguminous grasses are especially valuable in this respect. Protein productivity of legumes is 2-3 higher than that of cereals and amounts to 2-3 tonnes/ha of protein per vegetation due to symbiotic nitrogen fixation.

Perennial grasses effectively protect the soil from erosion, also in early spring and late autumn. By leaving a lot of organic matter, they cultivate low fertility soils. With the help of these crops, phytomelioration can be carried out, using, for example, milkvetch on saline soils, lupine - on acidic soils.

The cost of 1 k.u. and 1 kg of digestible protein of perennial grasses is several times lower than that of annual grasses. Cultivation costs are significantly reduced as there is no need for annual tillage and sowing.

Perennial grasses are used for hay, haylage, grass meal, silage, green mass and as a pasture crop. Some of them (sainfoin, milkvetch) are good mellifers.

Feed value. One kilogram of hay of perennial grasses contains on average 0.5 k.u., from 16 to 21% of crude protein in the dry matter of leguminous grasses and 9-14% of protein in the dry matter of cereal grasses. The most protein-rich species are alfalfa variegated (21%), creeping clover (21%), oriental goatgrass (20%), milkvetch (20%), hornwort (18%), sainfoin (16%), meadow clover (16%). In cereal grasses, protein content decreases to 14% in awnless bromegrass and meadow timothy, 13 in meadow fescue, 11 in wheatgrass, and 9% in wheatgrass. Leguminous grasses are characterised by a high content of essential amino acids (50-70 g per 1 kg of dry matter). Green mass is a good vitamin fodder. Alfalfa is especially valuable in this respect. It is a multivitamin fodder. In 1 kg of green mass of alfalfa contains 50 mg of carotene (provitamin A), 20 mg of vitamin C, 5 mg of vitamin B.

Leguminous grasses have high digestibility (63-75%), whereas cereals have 52-61% digestibility. Crops such as clover, lucerne, hedgehog, fescue have excellent digestibility. The digestibility of wheatgrass, wheatgrass, bromegrass and sainfoin is good, while that of melilot is satisfactory, as it has a high content of coumarin (an organic aromatic substance with a specific odour). There is less coumarin in plants in the evening and morning, as well as in the budding phase. Eatability of melilot increases when fed in a mixture with cereals and when ensiled.

Perennial grasses are characterised by high productivity. The most highly productive are blue alfalfa, meadow clover, white milkvetch, sand sainfoin, meadow timothy, hedgehog, blue and rootless wheatgrass, and reed fescue. They can yield from 3-8 to 10 tonnes/ha of hay. Blue alfalfa, for example, due to its very high regrowth capacity after mowing in Central Asia under irrigation can yield 5 harvests and up to 20 tonnes/ha of dry matter.

Pink clover, yellow alfalfa, meadow fescue, ridge-leaved honeysuckle, and lom kokolosnyk sitnikovy are considered less productive (by 30 per cent).

The average yield of perennial grasses for hay remains quite low and amounts to 1.5 tonnes/ha in the region.

14.2. Botanical characterisation

When using perennial crops in production, it is very important to know the distinctive features of their seeds, leaves, inflorescences, fruits (Tables 37-39, Fig. 16-18).

Flowers of grasses of the legume family are moth-type, asexual. They consist of five petals (a sail, two oars and a boat) and five sepals. Cross-pollination by insects predominates.

Distinguishing features of seeds of forage grasses of the legume family

Perennial grasses of the legume family				
type	size, mm	mould	surface	colouring
Red clover Trifolium pretense L.	1,7-2,2	Heart-shaped, lopsided	Brilliant	Yellow and purple
Pink clover (hybrid, Swedish) Trifolium hybridum L.	1-1,25	Heart-shaped, regular	Same	Dark green, almost to black.
White clover (creeping) Trifolium repens L.	1-1,25	Same	Same	Yellow, brown
Alfalfa (blue) Medicado sativa L.	2,2-2,5	Kidney-shaped, less often heart-shaped	Matte	Greyish-yellow, light brown.
Alfalfa yellow (sickle) Medicago falcate	1,75-2	Heart-shaped lobed	Same	greyish yellow
Alfalfa variegated Medicago varia	2,2-2,5	Kidney-shaped	Same	greyish yellow
Sainfoin viciafolia Onobrychis viciafolia Sc.	6-7	Weakly lobed	Smooth	Greenish-brown
Sand sainfoin Onobrychis arenaria D.C.	6-7	Same	Same	Same
White Melilotus albus	1,7-1,9	Sickle-shaped	Matte, less	Greenish-

Desr.		with a rubbing overhang	often slightly shiny	yellow
Melilotus officinalis Desr.	Same	Same	Same	Same
Oriental goatgrass Galega orientalis Lam.	3-5	Kidney-shaped	Matte	Greenish-yellow
Horned fritillary Lotus comiculatus L.	1,1-1,4	Slightly lobed, rounded.	Matte	Brown, less often green

Distinguishing features of perennial grasses of the legume family

View	Sheet type	Leaves, stem	Inflorescence	Flowers	Fruits
1	2	3	4	5	6
Meadow clover	triple	Ovoid with triangular pattern, not serrated, on short stalk, stem straight	The head is globular or oval at the top of the stems	Red, red-purple sessile flowers	1-2-seeded beans with a rounded ovoid shape
Clover pink	triple	Rhombic, oval, serrate, on short stalk, stem erect	The head is globular or oval, emerging from the leaf axils	Pale pink, flowers on pedicels	Same
White clover	triple	Ovoid with triangular pattern, serrate, on short pedicel, stem stalk stalked	The head is globular, emerging from the axils of the leaves	White flowers on pedicels	Same
Alfalfa blue	triple	Elliptic, obovate, with protruding vein, notched apex, middle leaf on a longer pedicel, bush erect	The brush is short, thick, dense, cylindrical	Purple and dark purple	Spiral-zipped bob, 2.5-4 turns

1	2	3	4	5	6
Alfalfa yellow	triple	Elongate-elliptic, narrowly lanceolate, with protruding vein, notched apex, middle leaf on a longer pedicel, bush collapsed	The brush is short, thick, dense, head-shaped	Yellows	The bob is sickle-shaped or straight
Alfalfa	triple	Elliptic to	The brush	Light violet,	Bob spirally

variegated		lanceolate, with notched apex, with a protruding vein, middle leaflet on a longer pedicel, bush raised	is short, thick, dense, cylindrical or head-like,	lilac, yellow, different colours (blue, yellow) occur	twisted, 2-3 turns
Sainfoin	Leafless	Lanceolate, with acuminate apex, entire margins	The brush is long, slender, loose, thinly pointed towards the apex	Pink, small, on short pedicels, sail shorter than or equal to the boat.	Rounded bob with firmly joined flaps, 4.5-5 mm, with short teeth

1	2	3	4	5	6
Sainfoin vicolabial	Leafless	Elliptical, full-edged	The ovoid brush is dense	Pink, on short pedicels, the sail is longer than the boat	Rounded bob with medium-sized teeth (0.5-1 mm)
White milkvetch	triple	Broadly oval, sparsely serrate along margins, with protruding vein, middle leaflet on a longer pedicel	The brush is long, thin	White flowers on short pedicels	Bob elliptic elliptic reticulate wrinkled surface
Donna yellow	triple	Rounded-ovate, serrate along margins, with protruding vein, middle leaflet on a longer pedicel	The brush is long, thin	Yellow, on short pedicels	Ovoid-shaped, transversely wrinkled bean
Oriental goatgrass	Leafless	Elliptical	Brush	Blue-violet	The bean is non-cracking
Horned loquat	triple	Obovate, irregularly rhomboidal, lanceolate, small, serrate on identical pedicels	Umbrella-shaped head of 5-8 flowers	Small, bright yellow	A multi-seeded bean that cracks at maturity

Distinguishing features of seeds of forage grasses of the bluegrass family

View	Value, MM	Shape	Sturgeonek	Awns or osseous	Flower scales

				pointing.	
1	2	3	4	5	6
Timofeevk a meadow Phleum pratense L.	1,5-1,75	ovoid	Absent	Absent ets	Non-reflective, silver.
Meadow fescue Festuca pratense Huds.	6-7	Lancet	Straight, 2 mm long	Absent	Coarse, inner scales boat-shaped, greenish grey
Pasture ryegrass Lolium perenne L.	5,5-6,5	Lancet	Flat, wide at the top, 1.5-2 mm	Absent	Same
Awnless bromegrass Bromas inennis L.	9-12	Broad lanceolate	Straight, round, bevelled, 3 mm	Absent	Outer scales are broad at the top, colour dark grey, less often greenish, purple
Hedgehog Dactylis glomerata L.	5-7	Triangular	Straight, round, 1 mm long	1 mm osseous pointing	Outer scales with keel, colour light yellow
Rootless wheatgrass Roegneria trachycaulon Nevski.	8-11	Lancet	Protruding, wide at the top, 1 mm	2 to 3 mm osseous acuminations	Outer scales without keel, light yellow colouration

1	2	3	4	5	6
Agropyron cri statura Gaerth.	5-6	Lancet	Protruding, wide at the top with a fossa	3 to 4 mm osseous acuminations	Outer scales goose-feathering light yellow
Agropyron pectinifonne Agropyron pectinifonne R.	5-6	Lancet	Same	Same	Light yellow
Desert ryegrass Agropyron desertorum R.	4-5	Lancet	Same	3 mm osseous cusps.	Pale green
Siberian ryegrass Agropyron sibiricum R.B.	4-5	Lancet	Same	1 mm osseous acuminations	Pale green
Ryegrass tall Arrhenaterum elatius M.etK.	8-10	Lancet	-	Awns from base of scales cranked 15-20 mm	Scales at base with long hairs, light yellow-green.

Ryegrass multiflorum Lolium multiflorum Lam.	6-6,5	Lancet	Flat, wide at the top, 1.5-2 mm	Awn at the top of the scales, 5-6 mm	Inner scales ciliate along the edges, scales greenish-grey
Elyinus sibiriens L. sibiriensis L.	5-12	Lancet	-	Awns at the top of the scales up to 25 mm long	Inner scales without cilia, scales greenish grey

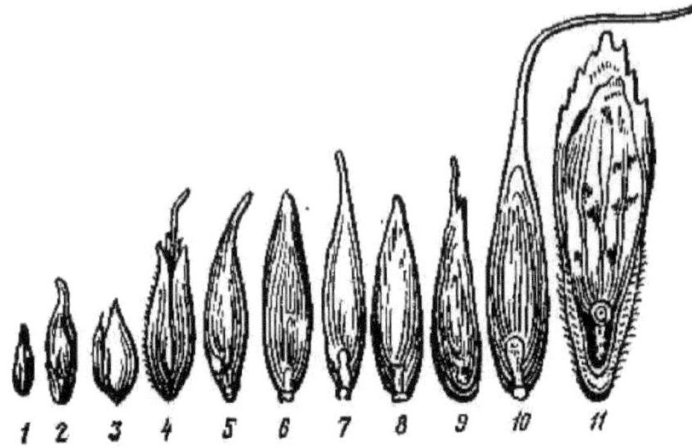

Figure 16. Seeds of perennial bluegrasses:
1 - white fieldwort; 2 - meadow bluegrass; 3 - meadow timothy meadow grass; 4 - meadow foxtail; 5 - hedgehog; 6 - meadow fescue; 7 - rootless wheatgrass; 8 - pasture ryegrass; 9 - crested wheatgrass; 10 - Siberian hairgrass; 11 - awnless bromegrass

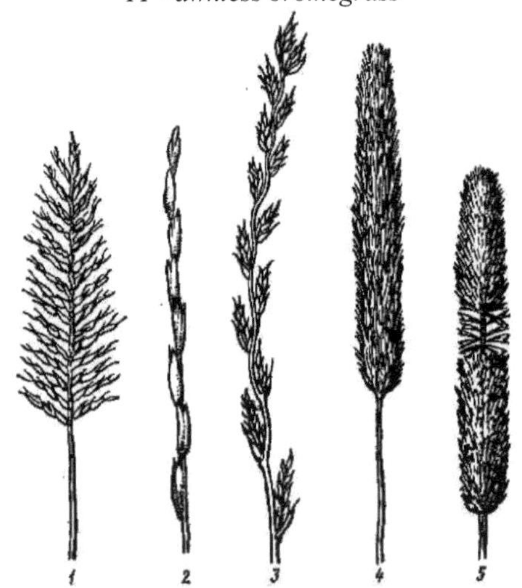

Figure 17: Inflorescences:
1 - spikelet of wheatgrass; 2 - spikelet of rootless wheatgrass; 3 - spikelet of

pasture ryegrass; 4 - spike panicle of meadow foxtail; 5 - spikelet panicle meadow timothy

Fig. 18. Inflorescences (panicles):
1 - bluegrass; 2 - white glade; 3 - awnless brome; 4 - meadow fescue; 5 - hedgerow

14.3. Biological characteristics perennial forage grasses

The main feature of perennial grasses is the ability to regrow after overwintering or mowing due to the reserve of plastic substances and buds laid in the regeneration zone. For cereals this is the tillering node or rhizome, and for legumes - the root neck (crown) - the transitional part between the root and the stem. In the first year, plants produce up to three shoots, in the second year - 15-17, in the third year - 20 and more. Each shoot lives for one year (or until mowing), in winter - dies off, but the root system and the zone of renewal are preserved. The winter resistance of the crop depends on the depth of this zone. For example, in clover the root neck is close to the soil surface, in the first year - at a depth of 1 cm, in subsequent years - at a depth of 4 cm. In blue and blue alfalfa - 7-10 cm, in yellow alfalfa - it sinks 28 cm with age. Therefore, clover is less winter-hardy than alfalfa and turfgrass.

Grasses grow to a height of 5-6 cm, so grasses should not be mown lower than 5-6 cm. Generative buds in the growth cone are formed slightly higher than vegetative buds, so before harvesting for seed, it is better to mow grasses at a height of 10 cm so that generative shoots are formed to a greater extent.

In spring and after mowing, regrowth occurs at the expense of nutrient reserves. For better overwintering and higher preservation of plants, a large reserve of plastic substances is necessary. The higher the concentration of cell sap due to the high sucrose content (up to 35% of ASV), the lower the freezing point of cell sap, the more resistant the plants are to frost. In order for plants to have time to undergo the hardening process and set buds of renewal, the last mowing should be carried out no later than 30 days before the onset of stable cold weather. In Siberia, this is the end of August. For better overwintering, the root rosette should have at least 5-6 leaves.

Of the nutritional elements, phosphorus and potassium above all increase the winter hardiness of the plants, whereas predominant nitrogen nutrition reduces winter hardiness. Grasses overgrown in autumn should be mowed to prevent them from dying out.

Heat requirements. Seeds of most perennial grasses germinate at a minimum temperature of +1...+20C, viable sprouts appear at +5...+6, optimal temperature for this period is +15...+200C. Sprouts tolerate frosts -6^{0}C. Optimal for growth is +20.+25^{0}C. In autumn, growth and development stops at temperatures below +5^{0}C, and in spring, growth resumes at +5^{0}C. The sum of active temperatures for the formation of the first crop for hay is 800-950^{0}C, and for the second crop - 600-800, for seed production - 1500-2100^{0}C.

The ability to withstand low winter temperatures is of great importance for perennial crops. The most frost-resistant and winter-hardy are yellow alfalfa, yellow-hybrid, variegated alfalfa, Siberian sainfoin, sainfoin yellow, blue alfalfa, sand sainfoin, white sainfoin. Poor winter hardiness is observed in meadow clover and pink clover. Meadow clover withstands in the first year of life up to -15^{0}C, in the second year at this temperature is thinned by half. Sainfoin and lucerne withstand up to -20.-30^{0}C with a good snow cover of at least 20 cm.

Perennial grasses of the bluegrass family are more winter-hardy, however, hedgehog has insufficiently high winter hardiness.

Moisture requirements. Grasses of the legume family tend to be more water-loving. They require up to 120% of the seed weight in water during germination. The critical period in relation to moisture is the budding phase and the regrowth period after mowing. The optimum soil moisture is 80-60% of NV. In seed production, a more moderate moisture content is necessary to prevent plant overgrowth to the detriment of seed production. Water consumption of perennial crops is high. The transpiration coefficient is from 800 to 1500.

Grasses with high drought tolerance include wheatgrass, wheatgrass, bromegrass, bromegrass, bromegrass, sainfoin, alfalfa, especially yellow and yellow-hybrid and variegated alfalfa. Alfalfa, for example, is drought-resistant due to its powerful, well-branched root system of up to 3 metres or more, capable of extracting water from deep soil horizons.

Clover, Lyadvenets, goatgrass, timofeevka, foxtail, bluegrass, ryegrass are considered more moisture-loving.

The most salt-tolerant soils are considered to be brittlebush, honeysuckle, turfgrass, alfalfa. On acidic soils (pH 4.5-5.5) perennial lupine, pink clover, horned clover, meadow timothy, meadow thymopheevka, and melilot grow relatively well.

Grasses of legume family are more demanding for such nutrition elements as phosphorus, potassium, molybdenum, boron, and yield of bluegrasses is more dependent on nitrogen.

All these crops are considered to be long-day, light-loving plants. When sown under the cover, Lyadvenets, clover are better able to withstand shading and somewhat worse - alfalfa, sainfoin.

Taking into account the biological features of forage grasses, it is necessary to select the right species composition and zoning of grasses when growing in certain soil and climatic zones. In general, it is necessary to give preference to grasses of legume family as more valuable crops with higher protein content, enriching the soil with nitrogen. It is recommended to grow clover in taiga and sub-taiga zones as a more moisture-reliant zone

meadow, on acidic soils - pink clover, horned clover, on open drylands and southern slopes of sub-taiga - alfalfa blue-hybrid, alfalfa variegated hybrid, melilot, bromegrass, hedgehog, in depressions - meadow timothy. In the southern forest-steppe - alfalfa blue-hybrid, pestrihybrid, sainfoin, melil, meadow fescue, bromegrass, rootless wheatgrass. In steppe it is necessary to choose more drought-resistant crops - lucerne variegated-rhybrid, yellow-hybrid, Hungarian sainfoin, sainfoin, turfgrass, meadow fescue, bromegrass, brittle loosestrife. On irrigation it is better to use blue alfalfa in pure form as a crop that gives the maximum number of cuts.

14.4. Cultivation technology for fodder purposes

Place in crop rotation. Perennial grasses are included in the main rotation of the crop rotation with a period of use of at least 2 years, they are also cultivated in withdrawal fields from 7 to 10 years in one field. Good forerunners for perennial grasses are winter and spring cereals, fallow crops, row crops, except beetroot, annual forage grasses. It is economically favourable to place seedbeds on fallow.

Tillage. Most species of perennial grasses develop a powerful root system from 1 to 3 metres, so the main tillage is done from autumn to a depth of 25-30 cm, mouldboard or no-tillage - for erosion-prone areas.

Ploughing with pre-stubble tillage after stubble predecessors is more qualitative.

In no-tillage, a higher stubble (15-18 cm) will encourage more snow accumulation. On sloping fields, slitting across the slope is effective for better moisture accumulation. Early spring harrowing is necessary to conserve moisture in spring. Shallow (3-4 cm) cultivation before sowing should be carried out with cultivators with flat tines, as they create a dense sole, cut weeds well, and do not turn the moistened soil layer to the surface.

Perennial grasses have very small seeds, so the soil must be well levelled. Pre-sowing cultivation is combined with harrowing and rolling.

Sowing. Perennial forage crops are often sown under the cover of annual crops because perennial grasses are slow growing and have low productivity in the year of sowing. In the first year the cover crop produces a full crop, while perennial grasses produce from the second year onwards. The advantage of cover crops is that slow-growing grasses are not able to resist weeds, and they are less likely to be weeded under cover. The stubble of a cover crop holds snow better. But from the point of view of development biology, grasses under cover lack light, water and nutrition elements, so in spring of the next year they grow worse and are more thinned compared to non-covered crops. In order to minimise these negative effects of cover crops, it is necessary to choose the right cover crop, which should minimally shade perennial grasses. In this sense, winter crops are worse than spring crops, as they are more bushy, often lodge, and shade the grasses. Among spring crops oats as a cover crop can be somewhat worse than wheat, barley. This is due to the fact that oats are more bushy, their leaves die off later, and in wet autumns they can regrow again. Therefore, grasses are more depressed under the cover of oats.

The cover crop should be harvested early, so that the grasses, coming out from under the cover, have time to develop sufficiently and accumulate nutrients for successful overwintering. Legume-oat mixtures are harvested early for green mass. They can be used as a cover for early sowing of perennial grasses. And for late sowing, millet-like crops for green mass (fodder millet, Sudan grass) can be used as a cover. Millet-like crops, having a slow rate of development in the initial phases, less oppress perennial grasses at the beginning of vegetation. Perennial grasses are less oppressed if they are sown in wide rows of cover crop sown in rows with row spacing of 30 cm.

Seeding rate of the cover crop should be reduced by 20-30%. Nitrogen fertiliser must not be applied.

In steppe areas the best results are obtained by coverless sowing of perennial grasses. In areas of sufficient and moderate moisture, as well as under irrigation, compacted sowing is possible. For example, after harvesting winter rye or pea-oat mixture for green mass in June, disking, rolling and sowing of alfalfa are carried out.

Seeds of perennial grasses are checked for germination before sowing. Seeds that do not germinate change colour. Yellow and brown alfalfa and clover seeds have reduced germination. If in a batch of legume grass seeds more than 20% of hard seeds, seeds are scarified using scarifiers, grinders, clover grinders not earlier than 1-2 weeks before sowing. Effective is air-thermal heating of seeds in the sun for 5-7 days or in dryers at temperatures up to +40^{0}C. To prevent diseases, seeds should be treated with systemic preparations 20-30 days before sowing. For more effective symbiotic nitrogen fixation, legume grass seeds should be inoculated, especially if the crop is sown in the field for the first time. Preparations used for this purpose (nitragyn, rhizotor-fin) contain strains of nodule bacteria specific for each crop. 200 g of the preparation is used per 1 kg of seeds. Molybdenisation of seeds is also effective, especially when sowing on acidic soils. Consumption of the preparation - 70 g of molybdenum-acidic ammonium per 1 kg of seeds. When sowing on neutral soils use 40 g/tonne of boric acid seeds. Treatment with bacterial preparations is carried out on the day of sowing, without access to sunlight, not combining with dressing.

When selecting sowing dates, it is important that the seeds get into a moist soil layer. In sub-taiga and forest-steppe zones it is better to sow perennial grasses in early spring under the cover of early harvested crops. In steppe areas, where the soil dries out early in spring, summer sowing dates can be used, timing the sowing to coincide with mass precipitation, but sowing without cover.

To obtain a high yield of perennial grasses for fodder purposes, the plant stand density should be 2-3 million/ha. To obtain a full grass stand, 5-6 million/ha of seeds are sown in the steppe, 6-7 million/ha in the forest-steppe and up to 7-8 million/ha in the foothills and on irrigation. The reasons for 3-4-fold increase in the number of sown seeds are low field germination of small-seeded crops, poor pre-sowing soil preparation, and uneven depth of seed placement. Small grass seeds (with the weight of 1000 seeds 1.5-2 g) germinate well from the depth of 1 -2 cm. Half of the seeds do not sprout from a depth of 3 cm, and single seeds sprout from a depth of 4 cm. Often when sowing in unleveled, unrolled soil using disc coulters seeds fall into the soil layer 0-7 cm and only 20% of them sprout. Therefore, the seeding rate is significantly increased (Table 40).

Table 40

Seeding rate of seeds of perennial grasses
at 100 per cent sowing capacity

Types	Seeding rate, kg/ha		Purity, %	Germination is present,%
	with scattered sowing	with row sowing	iya semesters	lyan- arnye
Meadow clover	15	14	92	75
Clover pink	10	9	92	70
White clover	10	9	88	70

Alfalfa blue	14	12	92	80
Alfalfa yellow	10	8	92	70
White milkvetch	22	18	94	80
Sainfoin	90	80	96	75
Horned loquat	12	10	90	75
Timothy meadow	12	10	90	75
Meadow fescue	25	18	92	80
A hedgehog	20	18	90	70
Tall ryegrass	28	20	95	75
Ryegrass	25	18	92	75
Perennial ryegrass	28	20	92	80
awnless bromegrass	28	22	92	75
Crested ryegrass	12	10	95	80
Kentucky bluegrass	13	13	85	60
White volesia	11	10	80	75
Meadow foxtail	20	16	80	70
Siberian wolfsbane	25	20	92	75
Rootless wheatgrass	20	16	92	75

Significantly increase the seeding rate also because during the overwintering period the plants are thinning out, especially in case of poor agrotechnics. Seed sowing depth is 2-3 cm in light soils, 0.5-1 cm in heavy soils. Seeds of cover crop are sown 5-6 cm deep.

The method of sowing perennial grasses is of great importance. It is important that grass seeds do not fall in the same row with the seeds of the cover crop, so it is better to sow inter-row sowing method, using grain-grass seeders (SZT-3,6), in which sowing of cover and perennial crops is carried out from different boxes and coulters at alternation of rows of cover crop and grass in 7.5 cm. In the absence of such seeders can be sown crosswise: first cover crop - to a depth of 6-7 cm, and then on the rolled soil - grasses to a depth of 1-2 cm. When sowing grasses under winter crops in spring, disc seeders are used across the winter crops, then harrowing is carried out. The best lighting regime of the cover crop is provided when the direction of the rows of the cover crop from north to south, and grasses - perpendicular. The cover crop is also sown in a scattered-row method using conventional grain drills with fertiliser sowing machines. Cover crop is sown in rows, and grass seeds are scattered between the rows through seed pipes removed from the coulters, the soil is rolled. To maintain the seeding rate of grasses when sowing with grain seeders, seeds are mixed on the day of sowing with ballast, using for this purpose sieved through a sieve superphosphate 20-25 kg / ha for cereals and 10-15 kg / ha - for legumes.

It is possible to sow under winter cover at the end of spring snow melt on frozen soil in a spreading method. The seeds settle in the upper soil layer after snow melt to an optimum depth of 1 cm.

Cover crop and grass seed should not be mixed in the same seed box, as leguminous grasses have very friable seeds and cereals have the opposite.

When sowing in the summer, mustard clumps can be sown 8-12 metres across the prevailing winds. It is also possible to use scattered clumps by sowing 300-400 g/ha of mustard, mixing it with a hectare rate of grass and sowing together.

Fertiliser. Perennial grasses respond very well to fertiliser. Legumes under good conditions for nitrogen fixation respond less to nitrogen fertilisers and are more demanding for phosphorus and potassium. Phosphorus and potassium are applied under the main tillage from autumn at 60 kg d.v. / ha for each element in the subtaiga zone, in the forest-steppe and steppe - phosphorus 60 kg d.v. / ha. When sowing in the row is effective phosphorus fertiliser 10-15 kg d.v./ha. If fertiliser has not been applied to the stock, top dressing in the second and subsequent years in early spring is effective. Leguminous grasses are better to feed phosphorus and potassium fertilisers 30-40 kg d.v./ha, but not in a scattered way, and cutting them into the turf grasses flat cutters - fertilisers. Cereals can be fed either full mineral or nitrogen fertiliser also at 30-40 kg d.v./ha. Fertilise grass mixtures, taking into account the proportion of components. If the legume component predominates (more than 50%), then in order not to suppress nitrogen-fixing activity of nodule bacteria, fertiliser should be applied as for leguminous grasses. If the cereal component predominates, fertilise as for cereal grasses.

Crop care includes the following: rolling before and after sowing; destruction of soil crust by rotary organs; pre-emergence harrowing with light harrows; timely harvesting of the cover crop at a high cut (15-20 cm); when harvesting for grain, straw should be harvested immediately; top dressing and harrowing after cutting; mowing of weeds; cutting of gaps in late autumn; reseeding of severely thinned grasses in spring.

Rejuvenation of alfalfa and koster grass stands is also carried out starting from the third year of life, using discing to a depth of 4-5 cm with an angle of attack of up to 25 degrees. In this case, partial cutting of the root neck is allowed, which leads to the initiation of germination of additional renewal buds, the emergence of new shoots. This technique is effective when the soil is moist and there are enough nutrition elements in it. Snow retention is carried out by compacting snow with ring rollers.

Harvesting and harvesting of fodder. The optimum cutting height for fodder purposes is 5-6 cm, and for high-stemmed grasses (e.g. turfgrass) - 12-14 cm. A higher cutting height of 8-10 cm is recommended in the first year of life, as well as if the grass is to be harvested for seeds the following year.

When harvesting for hay, haylage, silage, perennial grasses are cut in the phase of budding and flowering. For green fodder and grass meal, it is better to cut leguminous grasses at the beginning of budding, when the plants have the maximum amount of nutrients and carotene. From branching to full flowering and later, legume grasses increase their fibre content from 15 to 35%, while protein and carotene content decreases from 28 to 17% and from 478 to 273 mg/kg dry matter, respectively. In addition, during the flowering phase, the proportion of stems increases and the proportion of leaves decreases, while leaves have 23 times more protein. Nutrients are mainly in the flowers, which are more likely to fall off when harvested than the leaves. Closer to flowering, diseases (powdery mildew, brown rust) are more evident on the plants, and the quality of fodder deteriorates. If the first mowing is delayed, the plants grow back worse and the yield from the second mowing is significantly reduced.

Grass mixtures are mown no later than the beginning of flowering of the dominant component.

The yield of grasses decreases over time, as often grasses go into winter with a reduced reserve of nutrients, there is no insemination, so it is better to use grasses in hay rotation. For example, 4-year single-mowing rotation consists in dividing the field into 4 equal plots, and harvesting on these plots is carried out according to the following scheme: The 1st is mowed after insemination, the 2nd - before flowering, the 3rd - at flowering, the 4th - at flowering. In the following years the following scheme is followed (Table 41).

Table 41

Hay rotation scheme

Years	Plots			
Year 1	1	2	3	4
Year 2	2	3	4	1
Year 3	3	4	1	2
Year 4	4	1	2	3

There are many different ways of harvesting forage from seedy grasses.

Scattered hay. Grasses are mown, dried in swaths, raked and stacked. Hay is stacked at a moisture content of 17%. At such humidity, if you make 20-30 turns from a bundle of plants, they half slowly unwind, a small part of the plants will break. When laying over-dried hay with a moisture content of 15%, there are large losses from shattering, and the bundle of plants is fully unwound, the plants will break. Wet hay (18-20%) does not store well.

When harvesting loose hay, 35-40% of nutrients and carotene are lost, as drying takes place in the sun for a long time.

Mowing grasses with simultaneous conditioning promotes faster drying by 1.5-2 times, as the integrity of tissues is broken and moisture evaporation is faster. This is especially true for legume grasses, which are slower drying than cereals.

The best quality hay is obtained by active ventilation. The cut plants, which have been drying in the field up to 45% moisture content, are picked up and stacked on the stationary site in stacks, in the base of which lattice ducts with air ventilation are installed. A 2 m layer of hay is placed on the lattices for 3-5 days, then up to 6 m more.

Pressed hay is of higher quality, as it preserves leaves and flowers better, transport is simplified, and labour costs are reduced by 2-3 times. Hay dried to 30% moisture content is picked up by balers, pressed into bales, rolls, stacked in stables and dried. All types of hay should be made from legume-grass grass mixtures or from cereal grasses.

It is better to prepare haylage from leguminous grasses in pure form. Grasses cut and wilted in the field to a moisture content of 45-55% are chopped, transported and tamped into haylage trenches. The activity of anaerobic putrefactive bacteria is prevented by drying the grasses to 50% moisture, at that the sucking power of microorganisms is equal to the water-holding power of colloids in the cells, and anaerobic bacteria do not multiply. The activity of aerobic mould fungi is prevented by intensified tamping and sealing of haylage storages. This method of fodder preparation is more expensive, so more nutritious and vitamin-rich leguminous grasses should be used in order to justify the high quality of fodder. Legume grasses have high losses during natural drying in the field for hay, they are not suitable for silage in pure form, because they do not have enough sugars necessary for acid formation during the decomposition of sugars. But for haying leguminous grasses are a good raw material, as haylage is not acidic (pH = 5.0), the protein in it is very well preserved. Good quality haylage is equal to green mass. 1 kg of legume grass haylage contains 0.3-0.4 k.u., 20% protein, 100-110 mg carotene.

Vitamin hay is made from leguminous grasses in the budding phase, dried without direct sunlight under a canopy.

Protein and carotene are best preserved in vitamin flour, as drying takes place very quickly (in 1.5 hours) at high temperature in special vitamin flour machines (AVM). Taking into account its high energy consumption (220 kg of liquid fuel per 1 tonne of flour), only leguminous grasses at the end of stemming and budding are used for its preparation. Plants are mown, chopped to 1-3 cm and flour is prepared on the AVM. 1 kg of grass meal contains 0.7-0.9 k.u., 20-25% protein, 250-300 mg of carotene. After 6 months of storage 50-75% of carotene is lost. To prevent this, grass meal is prepared with the addition of antioxidants, for example, santoquin 200 g per 1 tonne of meal. Storage should be airtight. Carotene is well preserved in an environment with high carbon dioxide content, so bags of flour are covered with freshly cut grass.

14.5. Advantages and formulation of grass mixtures

Cultivation of 2-4-component grass mixtures provides 14-25% increase in yield compared to single-species crops. This is due to the fact that grass mixtures more fully utilise sunlight, moisture, nutrients, as cereal and legume components have different architectonics in the underground and aboveground parts, different removal of nutrients from the soil. The cereal component, as more resistant to unfavourable conditions, contributes to more stable yields, especially in case of perennial use. Cereal grasses, having a low protein content, are less valuable in this respect than legumes, so it is better to use them in a mixture, and at the expense of the legume component, a protein-balanced forage is obtained. Leguminous grasses in pure form do not dry out well in cuttings and lose more leaves when making hay. Leguminous grasses in pure form are poorly silageable. Feeding large quantities of legumes fresh causes tympany in animals, whereas forage from grass mixtures does not provoke it. Grass mixtures compete more successfully with weeds than single-species crops, apparently due to the different development rates of the different components. In perennial use, grass mixtures give a more stable and higher yield, because when one of the components falls out, the other component replenishes the missing yield.

It is important to choose the right components when making grass mixtures.

1. They should minimally suppress each other. Usually cereals somewhat suppress legumes. In this sense, the best component for meadow clover is meadow timothy, because in the first year it grows slowly and the yield is formed mainly by clover. The following year, the clover thins out considerably and the share of timotheevka increases.

2. The components should be equally well adapted to the local environmental conditions. For example, on acidic soils, pink clover is more suitable for acid-tolerant timothy fescue. In southern areas, a mixture of clover with meadow fescue is more productive due to the higher drought tolerance of fescue compared to timothy, and in northern areas, clover with timothy is more productive due to the higher winter hardiness and acid tolerance of timothy. The best component for blue-

hybrid alfalfa is awnless bromegrass, but in dry conditions for yellow-hybrid alfalfa or sickle-shaped alfalfa it is ridgegrass, and on irrigation blue alfalfa gives 3 harvests, and the optimum component for it is multilocular ryegrass, which is not inferior to alfalfa in terms of yield.

3. Components should have the same rate of yield formation, simultaneous onset of harvest ripeness, but it is desirable that their critical periods for moisture and nutrients do not coincide.

Numerous varieties of meadow clover are divided into subspecies (types):

- late-maturing single-crop winter type, not flowering in the first year of life, but flowering only after overwintering;
- early maturing, double-cropped, spring type, which, when sown early without cover, flower in the first year.

In Western Siberia, late-maturing single-crop clover is widespread, as it is more winter-hardy. In meadow timothy meadow clover, the harvesting ripeness, i.e. flush, coincides with the beginning of flowering of single-crop winter type meadow clover, while in meadow fescue the flush comes 14 days earlier than in timothy meadow clover and coincides with the beginning of flowering of double-crop spring type meadow clover. In this regard, the optimum component for single-crop clover will be meadow timothy and for double-crop clover - meadow fescue.

The following rules should be observed when composing grass mixtures:

1) Grass mixtures should be composed of components that are well adapted to the conditions in which they will be grown and that produce high yields;

2) short-term grass mixtures (2-3 years of use) include 2-3 species, medium-term (4-6 years of use) - 3-5 species, long-term (7-10 years of use) - 5-7 species;

3) short-term grass mixtures include early maturing juvenile grasses that give maximum yield in the 2nd year of life (meadow clover, melilot, tall ryegrass, perennial ryegrass). Medium- and long-term grass mixtures include medium-longevity grasses that give maximum yield in the 2-3rd year of life (alfalfa, goatgrass, pink clover, fescue, timothy, hedgehog, Siberian hairgrass, wheatgrass) and long-longevity grasses that give maximum yield in the 3-4th year of life (creeping clover, awnless bromegrass, honeysuckle, foxtail, bluegrass).

In more humidified zones, under irrigation, the share of legume component is increased, and in dry conditions the share of cereals, as a more stable insurance component, is increased. For haying use, the share of legume component is increased, and for haying and pasture -

the share of cereals.

When calculating the seeding rate, double mixtures include 70-80% of the full seeding rate of each species in pure form, triple mixtures - 40-60%.

14.6. Peculiarities of growing perennial grasses for seeds

Currently, the need for seeds of perennial grasses is not fully satisfied. One of the reasons is the low yield of seedlings. In the best case grass seed production should be concentrated in specialised seed farms with appropriate material and technical base, which receive elite seeds from elite seed farms, multiply them and provide farms with grass seeds for forage purposes. Large farms can allocate specialised seed production units for seed growing. Forage crops can also be used for seed in the 2nd and 3rd year of use, when the grass stand is thinning, but the seed production of each plant may even increase.

Seed plots are placed on fallow in specialised crop rotations. It is necessary to sow one crop and variety each from the group of mutually infesting crops in the crop rotation in order to exclude overpollination and contamination by hard-to-separate impurities. Spatial isolation for legumes - 200 m, for cereals - 400 m.

When growing clover, alfalfa and other legumes for seed purposes, areas with higher relief, on southern slopes are chosen and located close to forest stakes, forest belts, so that the plants do not grow to the detriment of seed productivity under increased moisture availability, mature faster, better pollinated by insects.

It is better to sow grasses without cover for seed, although cover sowing is not excluded. With spring coverless sowing of spring-type perennial grasses, it is possible to obtain a seed crop in the year of sowing. Winter grasses do not start to flower until after overwintering and therefore do not produce a seed crop in the year of sowing. Winter-type grasses include hedgehog, fescue, foxtail, brome, honeysuckle, late-maturing meadow clover. Spring-type plants or double crops include lucerne, early-maturing meadow clover, timothy, ryegrass.

Wide-row sowing with row spacing of 45-60-70 cm allows, with good care, to obtain a higher seed yield than row sowing, especially of leguminous grasses. For example, the yield of alfalfa in row sowing is 0.14 tonnes per hectare, and in wide-row sowing - 0.56 tonnes per hectare (according to ANIISKhoz). Yield increases significantly due to the fact that plants under wide-row sowing branch better, the number of generative organs increases, lighting, moisture and nutrient supply, and attendance of pollinating insects improves. Plants are better ventilated, less diseased and damaged by pests. For cereal grasses wide-row crops have an advantage over row crops, provided that inter-row treatments are carried out.

Seeding rate for growing for seed in wide-row crops is reduced to 1.5-2 million/ha. SZT- 3.6, SON-2.8, SST-12 seeders are used for sowing.

Plant care is the same as in forage crops. On wide-row seed crops, inter-row cultivation is done with KRN-4.2. The first treatment - in early May, 2nd - in late May at a depth of 4-6 cm, 3rd - after harvesting seeds at a depth of 15-20 cm for moisture accumulation. Herbicides are used to control weeds, insecticides are used against pests in case of mass appearance of pests.

Peculiarities of pollination of perennial grasses. Legumes are cross-pollinated plants, pollinated by insects. But there are difficulties with pollination. Fruits are set only in 10-20% of flowers formed on the plant, so the yield of legume grass seeds is very low (0.5-1 kg/ha on average).

Cultivated honeybees do not actually pollinate alfalfa and clover. In alfalfa, the pistil is arranged in a column of nine fused and one free stamen. When pollinating insects open the flower to take nectar, they receive a powerful click, so smaller honeybees try to take nectar from the side, but no pollination occurs. But larger bumblebees, as well as wild bees - leafcutter or megachile bees - are not afraid of the click and pollinate alfalfa well. Wild bees build cells for offspring in tree holes, houses, and worm burrows. Currently, megahills are domesticated, bred with cassettes in specialised farms and used on alfalfa seed crops.

Clover is also poorly pollinated by honeybees because its plants have a deep calyx and the proboscis of the cultivated bee is not long enough. Bumblebees and wild bees have longer proboscises and are the main pollinators. In two-leaf spring early-maturing clover, seed yields are higher from the second cutting and in one-leaf late-maturing clover after mowing, because the later flowers have smaller calyxes and are better pollinated. After mowing, more active branching begins, inflorescences are formed on additional shoots, which increases the seed yield. In addition, flowering is more friendly, coincides with the mass summer of pollinators, the plants are less lodged and less affected by seed-eating. Mowing is done in early June when the first buds appear.

To increase the intensity of work of honey bees on clover, they are trained to smell clover. In each hive in the morning put 100 g of sugar syrup infused on clover flowers. At the beginning of flowering, 4 bee-families per 1 ha of seed crops are taken out. Attendance of flowers by bees increases 14 times, and the yield of seeds - 2-4 times. At the same time, it is recommended to use not chemical insecticides, but the

biological preparation Entobacterin, harmless to insect pollinators, in pest control.
It is recommended to locate legume grass seed crops no further than 100-150 m from the nesting sites of wild pollinators (woodlands, spikes, forest belts). Special crop rotations for leguminous grass seed crops are required, leaving buffer strips of 4-6 m between fields, which should not be ploughed during the period of crop rotation use.
Significantly increases the yield of leguminous grass seeds by fertilisation - spraying with microelements, especially boron, as the fruit setting increases. 0.05% solution of boric acid is used.
To increase seed yield of cross-pollinated and wind-pollinated cereal grasses (bromegrass, bromegrass), artificial additional pollination is used by dragging a rope through the flowering grass.
The maximum yield of grass seeds is in the 1-2nd year of use. Old-aged crops are more damaged by pests. Therefore, it is better to use crops by alternating mowing every other year for hay and for seeds.
Seeds of perennial leguminous grasses ripen sporadically, the grass stand has high humidity, the share of fruits from the total plant mass is very small (3-5%). It is very difficult to separate seeds from the total mass of plants, so harvesting is often carried out separately, starting to mow plants when 75-80% of fruits on alfalfa, turfgrass and clover plants and 40-50% on sainfoin plants turn brown. Mow plants into swaths with reaper ZhRB-4,2, mower E-301. Swaths are picked up and threshed by combines equipped with the device PST-10, 54-108 for wiping bobs. The set of the device includes a rubbing device, additional mesh screens, ring flaps of the fan inlet windows and replaceable sprockets of the spike auger drive. Bean threshing is carried out at a rigid 1200-1300 rpm. The clearance between the deck and the drum is kept to a minimum. Threshed seeds through the sieve go to the hopper, and unthreshed seeds go to the threshing device for repeated threshing. If dryers are available on the farm, it is better to harvest legume seeds directly, the mown plants are transported to the transport vehicle and subsequently dried and threshed at the stationary plant. This significantly reduces losses. Direct harvesting with pre-desiccation with Reglon is also being introduced.
Seeds of cereal grasses crumble easily after maturity.
They are harvested either directly or separately. Harvesting should be carried out at maximum cut. Grass seeds of the bluegrass family have increased floatability, therefore it is necessary to regulate the air flow of the fan and thoroughly seal the combine.
Before cleaning, the seed crop is passed through a clover-grater to wipe the seeds from the beans and leguminous grass stubbles, as well as to

break up unthreshed spikelets and increase the bulkiness of cereals. In the absence of clover grinders, the teddy is passed twice through the combine drum with raised decks.

The seeds are then cleaned, dried, secondary cleaning and sorting.

15. ANNUAL FORAGE GRASSES

Annual forage grasses are of great importance in improving the fodder base, as these plants give high yields of green mass in the year of sowing, have high nutritional value, and have a short vegetation period. Mowing ripeness comes in 45-60 days, which makes these crops indispensable in intercrops and in occupied fallow. They are used for hay, haylage, silage, green mass.

15.1. Annual grasses of the legume family

The most valuable of this group of plants are annual leguminous grasses. These are sown vetch, mossy vetch, field pea (pelyushka), seradella (Tables 42, 43). The dry mass of vetch contains up to 19% protein, pelyushka - 19-20, seradella - 15%. Carotene content in green mass reaches 56-80 mg/kg.

Table 42

Distinguishing features of seeds of annual leguminous grasses

View	Shape	Colouring	Surface	Magnitude, mm
Vicia sativa Vicia sativa L.	Globular, squashed.	Yellowish-brown to black, often patterned	With a sparkle.	4,5-5
Vicia villosa Roth.	Globular	Black with no pattern	Matte	3-4
Field pea Pisum arvense L,	Round, slightly angular, with indentations	Grey, brown, patterned.	Matte	4-7
Seradella Ornithopus nativus	Unmilled, barrel-shaped, flattened bean segment	Greenish grey	Longitudinally reticulate-wrinkled.	2,7-3

Table 43

Distinguishing features of annual leguminous grasses

View	Leaves	Leaflets	Middle vein of the leaflet	Inflorescence, flowers
Vika sowing	Parene peduncles with tendrils	oblong-lined, with obtuse apex, notched, slightly pubescent	Protrudes over the edge of the leaf	Flowers in the axils of leaves 2-3, purple-red
Vika furry	Same	Oval-oblong, strongly pubescent	Does not protrude beyond the edge of the leaf	Brush long, multiflowered, flowers violet-blue
Field peas	Pinnately pinnate with tendrils and a	Oval	Same	Flowers in leaf axils 2-3, red-purple each

	large bract with a red spot at its base			
Seradella	Leafless	Oblong-oval	Same	Small umbrella of 3-5 flowers, the flowers are pinkish-white in colour

Biological characteristics of leguminous grasses. They are long-day plants, so they are moderately demanding to heat. The minimum seed germination temperature is +2...+30C, the optimum temperature is +150C. Sprouts can withstand frosts down to -5...-60C. Optimum temperature for vegetative mass formation is +15...+160C, for seed ripening - +16.+20$^{(0)C}$. The sum of active temperatures for the formation of green mass - 900^{0}C (50-60 days), for the formation of seed yield - 1900^{0}C (90-100 days).
These plants are moisture-loving, so they are usually grown in more humid areas. The maximum need for moisture is during the budding - flowering - seed filling period.
Relation to soils. Seed vetch succeeds better on cohesive soils with good water-holding capacity, tolerates slightly acidic soils with pH = 5-6.5. Vika mossy is less demanding, it succeeds well on loam, sandy soils, so it is called sandy. But it is better that the soil contains a lot of calcium, so it does not tolerate acidic soils. Seradella can grow successfully on acidic soils.
Biological forms of vetch: sown vetch - spring vetch, mossy vetch - winter vetch. But mossy vetch also has spring forms of more southern origin. For example, in the Altai region there is a variety of spring type mossy vetch Nezhnostebelnaya of the ANIISKh selection, created on the basis of varieties selected in Turkmenistan, so it is characterised by increased drought resistance and salt tolerance.
Field peas are not demanding to the soil, tolerates weakly acidic sandy, peaty soils, but does not grow on wet, swampy.
Often, seed peas are grown for green mass due to lack of seeds in farms. However, it is better to use pelushka for this purpose, as it has a number of advantages in this respect: it is more drought-resistant than seed peas, as it is more small-seeded. It makes better use of rainfall in the second half of the summer, as it uses, as a rule, later maturing varieties, which have a higher yield of green mass. Seed consumption is lower and the multiplication rate is higher. When sown in a mixture with oats, it gives a higher yield, as it is less oppressed by oats than sown peas. It tolerates shading better, so it provides a higher protein content in the mixture. In addition, it is less demanding to soils, less affected by diseases and pests (aphids).

Cultivation technology. In field rotations these crops are sown in the fallow (for hay and green mass), in fodder rotations - after winter and spring crops, tilled crops. These crops themselves are good forerunners, as at early harvesting they clear the fields of weeds and deplete the soil of nitrogen less.

Sowing dates. For seeds sow early in early May, for green mass - in 2-3 terms with an interval of 15-20 days to organise a green conveyor. Seeding rates of vetch in mixture with oats: 2-2.5 million/ha of germinated vetch seeds (110-130 kg/ha) + 1.5-2 million/ha of oats (50-90 kg/ha).

Winter vetch is sown in a mixture with rye. But it is better not to sow them at the same time. Two weeks before sowing rye sow vetch, as it is slower to develop, and then - rye. Seeding rate: 3-4 million/ha of mossy vetch (75-100 kg/ha) + 1.5-2.5 million/ha of winter rye (60-80 kg/ha).

Harvesting for green mass can be done in early June if rye predominates in the crop and in July if vetch predominates.

Seeding rate for pellets: 1.2 million/ha of pellets (150-180 kg/ha) + 2 million/ha of oats (70 kg/ha).

Harvesting vetch for green mass should be done during the fruiting phase. This is due to the fact that vetch accumulates only 40-45% of dry matter yield in the flowering phase, and at the same time its stems coarsen slowly and little fibre is formed, unlike perennial grasses. Pelushka is also mown during the fruiting phase.

15.2. Annual grasses of the bluegrass family

The group of these plants includes Sudan grass, mogar, fodder millet. They are characterised by high productivity. Hay yield can be 5-10 tonnes/ha. Hay contains up to 9-10% protein. In 1 kg of green mass - up to 50-80 mg of carotene.

Biological characterisation. By origin these crops are short-day crops, therefore more demanding to heat and drought-resistant. Minimum germination temperature is +8...+10° C, optimum for growth is +25...+28° C. Sprouts die at frosts -3° C. The sum of active temperatures for full seed ripening is as follows
2000-2300^{0}C.

The rhythm of development of these plants coincides with precipitation in the steppe in the south of Western Siberia. At the beginning of vegetation they develop slowly and tolerate early drought well, and with July precipitation they start to develop rapidly and form a large biomass. These crops are most in demand in the steppe, where other fodder crops are less competitive.

Sudanka, millet and mogar grow well on chernozem and chestnut soils. Mogar is less demanding than Sudanka and can grow on sandy and sandy loam soils, as well as on heavy loams.

Cultivation technology. In crop rotation annual bluegrasses are placed after grain legumes, maize, winter crops.

These crops respond well to organic and mineral fertilisers, especially nitrogen fertilisers, as they, above all, increase the yield of green mass. $N_{45-60}P_{30-45}$ kg d.w/ha increase yields by 20-25%. Potassium is effective only on sandy, podzolised soils.

Tillage is similar to that for cereal crops.

Sowing dates are late, in late May - early June, when the soil warms up to $+10^0$C at a depth of 10 cm and the threat of frosts passes. Seeding rate: Sudanka in steppe - 10-15 kg/ha, in forest-steppe - 25-30; Mogar in steppe - 8-12, in forest-steppe - 15; fodder millet - up to 30-35 kg/ha. Seeding depth - 3-4 cm. The sowing method for fodder is row sowing, for seeds - wide-row sowing.

Harvesting for fodder is carried out in the flush phase at a height of 7-8 cm. These crops are fallow, regrow after cutting, and a second crop can be harvested if moisture and nutrients are well supplied.

BIBLIOGRAPHY

1. Agronomy: textbook / ed. by N.N. Tretyakov. Moscow: Academy, 2004. 480 c.
2. Barsukov A.I. Spring wheat in Kulunda / A.I. Barsukov. Barnaul: Alt. book publishing house, 1983. 104 c.
3. Cultivation of spring durum wheat in Altai Krai: recommendations of the Siberian Branch of the Russian Academy of Agricultural Sciences. Barnaul, 1999. 36 c.
4. Vedrov N.G. Practicum on crop production: textbook / N.G. Vedrov, E.T. Zagorodnaya, E.M. Nesterenko, I.N. Frolov. Krasnoyarsk: Izd-vo Krasnoyarsk University, 1992. 384 c.
5. Cultivation of sunflower by industrial technology: methodical recommendations. Novosibirsk, 1984. 45 c.
6. State catalogue of pesticides and agrochemicals permitted on the territory of the Russian Federation. 2007.
7. Agriculture in Siberia: textbook for universities / N.V. Yashutin, A.P. Drobyshev; ed. by N.V. Yashutin. Barnaul: Izd-vo AGAU, 2004. 414 c.
8. Leguminous crops / ed. by D. Shpaar. Minsk: FAUinform, 2000. 327 c.
9. Grain crops / ed. by D. Shpaar. Minsk: FAUinform, 2000. 342 c.
10. Ilyin V.S. Early maturing maize for grain in Western Siberia / V.S. Ilyin, V.S. Gatsenbiller. Barnaul, 1995. 158 c.
11. Korenev G.V. Plant growing with the basics of breeding and seed production / G.V. Korenev. Moscow: Agropromizdat, 1991. 569 c.
12. Malting barley in the Altai Krai: method. recommendations / ANIIZiS. Barnaul, 2003. 43 c.
13. Seed production of grasses in the Altai Krai: method. recommendations / ANIIZiS. Barnaul, 1984. 27 c.

APPENDICES.

Annex 1

Agricultural machinery

Mark	Deciphering
1	2
KPSH-9, KPSH-5 KPE-3,8 PG-3-5 GUN-4 KPG-250 KPP-2,2 PN-8-35 PKA-2A PON-2-30 PL-5 25 OPT-3-5 KPS-4 KD-6 LDG-15 BD-10 3BZS-1,0 3BP-6,0 3OR-0,7 SHB-2,5 BMSH-15 BSO-4 PAV-6 CVU-2,6 BIG-3 3KK-6A 3KKSH-6A KRN-4,2 USMK-5,4 CON-2.8 STN-2.8	Cultivator-plough Cutter broad tiller Cultivator erosion control Cutter-plough deep loosener Lubo loosener-fertiliser Deep loosening cultivator-plough tiller Trailed cultivator-plough tiller Eight-corner mounted plough Tillage combination machine Trailed reversible plough Five-hull plough plough Grass bed cultivator Trailed high-speed cultivator Disc cultivator Hydraulic disc harrow Disc harrow Three-section harrow medium tooth harrow Three-section seed harrow Three-section lightweight rayboronka Schleif harrow Wide-tine harrow Lightweight harrow Soil cultivation unit levelling machine Snowplough Roller Roller reinforced Hydrified needle harrow Three-section ringed roller Three-link ringed-spur roller Cultivator-plant feeder mounted Universal beet mechanical thinning machine Cultivator-cultivator-mounted Cultivator-cultivator-mounted fertiliser sower
1	2
RUM-8 RMG-4 AIR-20 AIC-12 PS-10 OPSH-15 SZS-2,1 SZP-3,6 PPK-12,4 SKS-3,6 SZU-3,6 SZT-3,6 SKPN-8 SUPN-8 SST-12 STV-12 СК-5 СКД-5 ЖВН-6	Fertiliser spreader Fertiliser spreader Fertiliser spreader Fertiliser shredding machine Mobile unit for preparation of working liquids Seed dressing Опрыскиватель прицепной штанговый Сеялка-культиватор зерновая стерневая Сеялка зерно-туковая прессовая Почвообрабатывающий посевной комплекс Сеялка-культиватор стерневая Сеялка зерно-туковая узкорядная Сеялка зерно-туковая травяная Сеялка кукурузная пунктирная навесная Сеялка универсальная

ППТ-3 PST KZS-20SH	пневматическая навесная Сеялка свекловичная туковая Сеялка точного высева Самоходный комбайн Self-propelled double-drum harvester Roller harvester mounted Pick-up web-transporter Harvester for harvesting grass seeds Grain cleaning drying complex

Annex 2

Classification of soils by availability of mobile phosphorus and exchangeable potassium, mg/kg of soil

Provisioning	P_2O_5			K_2O		
	according to Kirsanov, sour	Ma-chigin, carbonate	according to Chirikov, non-carbonate	according to Kirsanov, sour	Ma-chigin, carbonate	according to Chirikov, non-carbonate
Very low	Less 25	Less 10	Less 20	Less 40	Less 50	Less 20
Low	26-50	11-15	21-50	41-80	51-100	21-40
Medium	51-100	16-30	51-100	81-120	101-200	41-80
Elevated	101-150	31-45	101-150	121-170	201-300	81-120
High	151-250	46-60	151-200	171-250	301-400	121-180
Very high	More 250	More 60	More 200	More 250	More 400	More 180

Annex 3

Soil supply of easily hydrolysable nitrogen (mg/kg of soil) depending on soil pH

Provisioning	pH less than 5.0	pH 5-6	pH greater than 6
Very low	50	40	40
Low	70	60	50
Medium	70-100	60-80	60-70
High	More than 100	More than 80	More than 70

Annex 4

Removal of nutrients from 1 kg of main crop and relevant by-products of agricultural crops, kg

Culture	N	P2O5	K2O
Winter wheat	3,5	1,2	2,6
Winter rye	3,0	1,2	2,8
Spring wheat	3,8	1,2	2,5
Barley	2,7	1,1	2,4
Oats	3,0	1,3	2,9
Maize	3,4	1,2	3,7
Millet	3,3	1,0	3,4
Buckwheat	3,0	1,5	4,0
Peas	3,0	1,6	2,0
Sunflower	6,0	2,6	18
Long flax	8,0	4,0	7,0
Potatoes	0,6	0,2	0,9
Sugar beet	0,59	0,18	0,75
Peas + oats for green fodder	0,3	0,14	0,5
Maize for silage	0,25	0,12	0,45

Annex 5

Coefficients of utilisation of nutrient elements from soil and fertilisers

Culture	Utilisation rates from soil			Utilisation rates from mineral fertilisers		
	N	P	K	N	P	K
Winter rye	0,25	0,1	0,13	0,65	0,32	0,7
Spring wheat	0,3	0,12	0,1	0,7	0,35	0,68
Barley	0,2	0,07	0,08	0,55	0,24	0,65
Oats	0,3	0,13	0,1	0,7	0,4	0,8
Maize	0,35	0,15	0,33	0,85	0,35	0,95
Sugar beet	0,3	0,1	0,3	0,76	0,25	0,8
Potatoes	0,2	0,05	0,13	0,6	0,2	0,8

Printed by Books on Demand GmbH, Norderstedt / Germany